DATE DUE

OC 13 '95			

COSMOGRAPHY

BOOKS BY R. BUCKMINSTER FULLER

Nine Chains to the Moon

4D Timelock

Untitled Epic Poem on the History of Industrialization

Ideas and Integrities

No More Secondhand God

Education Automation

Utopia or Oblivion

Operating Manual for Spaceship Earth

Design Science Decade Documents

Intuition

Earth, Inc.

Buckminster Fuller to Children of Earth

Synergetics: Explorations in the Geometry of Thinking

Synergetics 2: Further Explorations in the Geometry of Thinking

Tetrascroll

And It Came to Pass—Not to Stay

Buckminster Fuller on Education

Critical Path

Grunch of Giants

Inventions

Synergetics Dictionary: The Mind of Buckminster Fuller

COSMOGRAPHY

R. BUCKMINSTER FULLER

Adjuvant: Kiyoshi Kuromiya

A POSTHUMOUS SCENARIO
FOR THE FUTURE OF HUMANITY

MACMILLAN
PUBLISHING
COMPANY
NEW YORK

MAXWELL
MACMILLAN
CANADA
TORONTO

MAXWELL
MACMILLAN
INTERNATIONAL
NEW YORK   OXFORD
SINGAPORE   SYDNEY

Macmillan Publishing Company
866 Third Avenue, New York, NY 10022

Maxwell Macmillan Canada, Inc.
1200 Eglinton Avenue East, Suite 200
Don Mills, Ontario M3C 3N1

Macmillan Publishing Company is part of the Maxwell Communication Group of Companies.

Library of Congress Cataloging-in-Publication Data
Fuller, R. Buckminster (Richard Buckminster), 1895–1983.
   Cosmography / R. Buckminster Fuller: adjuvant, Kiyoshi Kuromiya.
     p. cm.
   Includes index.
   ISBN 0-02-541850-5
   1. Cosmography. I. Kuromiya, Kiyoshi. II. Title.
GA9.F85   1992
113—dc20   91-33459   CIP

Macmillan books are available at special discounts for bulk purchases
for sales promotions, premiums, fund-raising, or educational use.
For details, contact:
Special Sales Director
Macmillan Publishing Company
866 Third Avenue
New York, NY 10022

10 9 8 7 6 5 4 3 2 1

PRINTED IN THE UNITED STATES OF AMERICA

CONTENTS

A NOTE FROM THE ADJUVANT   vii

1. The Dawn of Einstein's Universe   1

2. Discoveries of the Human Mind   32

3. Einstein   75

4. Historical Underpinnings   91

5. Taking Inventory   107

6. Cosmic Conceptioning   117

7. Integrity   248

INDEX   267

NOTE FROM THE ADJUVANT

ON THE DAY THAT R. BUCKMINSTER FULLER died at age 87 in 1983, I found the entire *Cosmography* manuscript, which we had been working on, neatly stacked in the middle of an uncharacteristically tidy desk in his study at his Pacific Palisades, California, home. Atop the manuscript was a note addressed to his daughter Allegra (Snyder), and his grandson Jaime and granddaughter Alexandra. It began, "If something happens to me and I [should] die suddenly, I want you to know of the extraordinary importance of my now being written book *Cosmography* . . ."

THIS BOOK, ITS ORGANIZATION AND CONTENTS, were conceived as a whole and nurtured through to virtual completion by Fuller during his last four years of life. All of the vocabulary and concepts originate in Fuller's mind, and the way they are phrased is his. I have served as adjuvant, a term Fuller borrowed from medicine (specifically immunology) in 1980 to designate my role in the writing of *Critical Path*—that of a "helper" in transcribing and editorially refining for publication his ideas, words, and extemporaneous "thinking out loud." In *Cosmography,* as in *Critical Path,* I have served in this Fuller-designated role to preserve the idiosyncratic concepts, tone, syntax, and phraseology of Fuller in preparing the manuscript for publication. If it has strayed at all from his original conception, the blame is mine. If it is a faithful representation of his methodology and thought, the credit goes to him and his insistent and uncompromising resoluteness. I have not attempted to flesh out those few parts of the book that Fuller left unfinished. I hope the reader will patiently bear with us, as many have during the eight years that have passed between Fuller's death and the publication of this long-awaited final book. Fuller would perhaps attribute this span of time to nature's own purposeful system of gestation rates.

—Kiyoshi Kuromiya
Adjuvant

COSMOGRAPHY

THE DAWN OF EINSTEIN'S UNIVERSE

THE DARK AGES STILL REIGN over all humanity, and the depth and persistence of this domination are only now becoming clear.

This Dark Ages prison has no steel bars, chains, or locks. Instead, it is locked by misorientation and built of misinformation. Caught up in a plethora of conditioned reflexes and driven by the human ego, both warden and prisoner attempt meagerly to compete with God. All are intractably skeptical of what they do not understand.

We are powerfully imprisoned in these Dark Ages simply by the terms in which we have been conditioned to think.

Some concepts have been long imagined by humans to be real: up and down, straight lines that extend to infinity, measurement based on squares and cubes. For ages, humans have mistakenly thought that solids were truly solid and that several lines could conceivably pass through the same point at the same time. Humans have deceived themselves that the existence of one, two, and three dimensions is independently demonstrable and that there is factual evidence proving the existence of more than one race of human beings. And further, humans attest to belief in God, although only paying ''him'' once-a-week lip service in an otherwise human, male-dominated Universe.

Formal religions have been organized to attend to the otherwise inconvenient, constant recognition of God, while humanity gives six-sevenths of its time to rendering service to the exclusively selfish dictates of human power structures.

Misorientation, wrong beliefs, and conditioned fixations are escapable only when that which is physically and metaphysically true becomes experimentally provable and comprehensible. The untrue is rendered spontaneously obsolete only by the demonstration of that which is true. It is here I am compelled to begin.

The only important fact about me, as I write this book, is that I am an average, healthy human being. There is nothing that I have done that could not have been done equally well or better by any other healthy human being, given the unique working circumstances under which I have operated for the last fifty-five years.

This book presents my individual efforts to escape from the clutches of the Dark Ages, but in a larger sense it shows the beginnings of our species' epochal rebirth, what I call the dawn of Einstein's Universe.

As an average, healthy human being, I have learned how little we know about ourselves. For an instance: Why are we humans included in the design of Universe? How is the designing of eternally regenerative Universe both anticipatorally and progressively conceived and realized, together with the part already played and as yet to be played by humans?

I am sure that the only reason that I am widely known is because in 1927, when I was thirty-two, I decided to make an experiment of myself. The experiment sought to discover and realize what would happen if a healthy, moneyless, unknown individual with dependent wife and newborn child altogether discarded the assumption that an honorable human must earn the right of family and self to live ("earn a living") and do so to the satisfaction of the socioeconomic power structure governing the political system in which he lived and, breaking away from all socially accepted concepts of the significance of human presence on planet Earth, undertook to discover what—if anything—a mature individual might be able to do effectively on behalf of all humanity that would be inherently impossible of accomplishment by any political system, nation, or private-enterprise corporation no matter how powerful or well-endowed.

Because (*a*) I had no competitors in such an initiative and (*b*) the experiment has been so richly productive, I have come to be widely known. If there had been any competitors, you would probably never have heard of me. If there had been competitors, I would long ago have dropped out, leaving the task to the competence of my competitors. I initiated that which I did only because I was convinced it needed to be done and to the best of my knowledge no one else was attempting to do it.

I was thirty-two years old. In the year I was born, the life insurance companies' actuarial tables showed that life expectancy was forty-two years for white males born in New England. Being thirty-two, it was my feeling that I had only ten years left within which to carry out my experiment. I realized even then that I would get nowhere by asking the three billion humans then on planet Earth to listen to me—let alone support me. As a rule, I found that people listen only when *they* ask you to speak to them.

It seemed clear to me that the only possible way I could become effective would be by doing what I did on a scale all out of proportion to what one would imagine possible for a mere individual. First, it would have to be done on behalf of all humanity, and second, it would have to take advantage of the human mind's capability to discover the generalized principles. These generalized principles govern the operation of our physical Universe. I set out to discover the entire inventory of generalized, only-mathematically-expressible scientific principles that had thus far in history been discovered by humans. Third, I surmised that I must employ those principles to develop artifacts that would render the living environment more favorable for all humans and their supportive ecology.

The role of human mind, invention, and tools in the relentless course of human cosmic evolution became quite clear to me.

My hope was that the development of this more favorable physical environment would bring about such a reduction of physical disadvantage to humanity that individuals with vastly greater knowledge of their technological options would become principally concerned with unselfish goals: the realization of potential advantages for all humanity to be attained only by an artifact revolution. Such developments would encourage worldwide understanding and social accords sufficient to entirely eliminate the local condition of degeneration or prolonged economic want and anguish. Ultimately, I hoped that competition for limited resources would be ended and thus the root cause for war.

Perhaps on some level, expressed or unexpressed, similar motivation drives all those who set out to discover and invent.

OVER A HALF CENTURY AGO, when I embarked on my "experiment in individual initiative," I set before myself (as I have repeatedly ever since then) one very large question: What is our human function here in Universe?

My first answer to that question came from three closely related observations:

1. That all the known living organisms other than humans have some integral bodily equipment that gives them special operating capability in special environments.
2. That many creatures, including humans, have brains and that brains are always and only sorting the information reported by the senses and integrating this information into system images and therewith coordinating nervous control responses or forming improved new system imaginings. Brains are therefore always dealing with special-case experience—for example, "This one smells a little sweeter than that one." Brains must sleep periodically. Brains deal in beginnings and endings of special-case considerations. Brains are physical, temporal, and frequently terminaled.
3. Humans also have a faculty unidentified with any other creatures—the faculty of *mind*. Minds are always and only concerned with the discovery of eternal, constant interrelationships manifest in a myriad of special-case experiences of the brain, which interrelationships are not to be found in any one of the special-case system components considered separately.

One of the most important events of classical science involving the interrelationship findings by the human mind is demonstrated by the mathematician-astronomer Johannes Kepler, whose story I shall recount here.

Based on his accurate observations and measurements, Kepler found that all the planets of which he was aware (*a*) were of different sizes, (*b*) operated at different distances from the Sun, (*c*) orbited the Sun at different rates, and (*d*) traveled their respective orbits at different rates. Kepler said that the planets, though apparently on the same team, seemed to be utterly disordered. He then said they did share one thing: the fact of all going around the same Sun. As a mathematician, he knew he could assign these planets something else in common. He also knew that given two known constants, one may discover other interrelationships within the team. Kepler then assigned a common constant to each and all the known planets—exactly the same increment of calendar time.

Starting at the same moment of calendar time and finishing at the same moment of calendar time, Kepler observed and recorded the planets' concurrent orbital travel over a twenty-one-day period. This gave him the data for graphing the slices-of-pie-shaped, triangular patterns formed by the starting and finishing radii of measured distance from the Sun to each planet at the start and finish of the twenty-one-day event. The arc of travel distance between the start and finish closed the radii ends to form triangular shapes. Kepler intuitively decided to calculate the area of each of them. Doing so, he found that they were not only similar areas but were elegantly, exactly the same size.

He surmised that the planets could not sweep out exactly the same cosmic areas unless they were coordinating in some exact manner. Since the planets were not touching one another, they could not be coordinating like toothed gears. Far from touching, these massive bodies were rotating and orbiting millions of miles distant from one another.

Kepler was forced to conclude that there was an invisible, unsmellable, soundless, untouchable, intertensionally restraining force governing the planets' orbital motions.

The work and findings of Kepler's contemporary Galileo regarding the exact mathematical rate of acceleration of "falling bodies" led to Isaac Newton's discovering the mathematical expression of the gravitation laws of Universe. Newton found that the interattraction of any two celestial bodies always varies inversely with the second power of the arithmetical distances intervening. Thus, halve the distance, and increase the interattractiveness fourfold.

Here again we have the human mind discovering what the brain's sensing is utterly incapable of apprehending. The mind can, and does, from time to time discover the only mathematically expressible laws governing these nonsensorially discoverable macro-microcosmic interrelationships which always hold true in all special-case instances. When such initial discoveries are found to be exceptionless, they become known as "laws"—hence, the generalized laws of science.

Exceptionlessness can be termed eternal. Human mind has discovered a meager inventory of these only mathematically statable, eternal laws governing the physical design and operation of Universe. These laws have never been found to contradict one another. All have been found to be interaccommodative. All of them may be objectively employed in special-case technology.

Humans possessed of the family of generalized mathematical laws governing all the relevant, variable factors in aerodynamics are able to build a flying machine by which they can outfly birds in speed and altitude. Humans can lend one another their "wings."

That humans alone of all known phenomena have access to the great design laws of Universe immediately implies that we must have been introduced into Universe for some very significant ultimate functioning.

I realized that humans must have been given their extraordinary minds in order to discover principles, the conceptual comprehension of which permits invention and development of instruments and tools. With these instruments and tools we can explore our immediate senses-apprehended environment as well as our vast outward macrocosmic instrument reachings and exquisite inward microcosmic penetratings of our locally experienced scenario Universe.

In 1923, E. P. Hubble discovered another galaxy. Between that event and November 1982, astronomers using the new radio telescopes were able to "see" through the great dust clouds of our Milky Way and discover a hundred billion additional galaxies. The accelerating rate in the increase of acquisition of ever more exact macro- and microinformation seems beyond comprehension. Only by such nonsensorially apprehended, macro-micro, experience-obtained information do we discover ever new challenges to our unique problem-solving capability as provided for by our eternal principles-discovering and -comprehending minds.

Another of the concepts leading to my discovery of a logical answer to why humans are included in the design of Universe is illustrated by the following example.

In the forward cockpit of the Boeing 747 and all other air transports there are an enormous number of computer-activated instruments. In flight, those instruments are constantly and exactly reporting all the thus-far-known-to-exist and knowable critical conditions operating throughout the airplane's power plant, airframe, landing gear, etc.

This suggested to me the following: that through all those instruments his cockpit team is monitoring, the captain of the Boeing 747 is apprehending in flight all of the airplane's locally critical (relevant) information, and through all the captain's total experience and his information-integrating brain and his physical-principle-comprehending mind, he is serving not only as the ship's comprehensive information harvester and integrator (teleologist) but also as the ship's constant comprehensive local problem-solver in maintenance of the integrity not only of the airplane and its passengers but also of local Universe and thereby of eternally regenerative Universe.

From this model, I made the following working assumption: Since it is only through our ability to use mind-discovered cosmic principles and the therefrom-developed instruments that we gain information, no matter how many more human-mind-performable cosmic-scale functions we may in time discover ourselves capable of coping with, it seems to be confirmed that the cosmic function of humans is indeed analogous to that of the captain of the Boeing 747, together with his pilots, engineers, and their galaxy of instruments; they, as we humans, are here for local-Universe information gathering and local-Universe problem solving in support of the integrity of eternally regenerative Universe.

What is common to all humans in all history is problems, problems, and more problems. If you are good at problem solving, you do not eventually arrive at Utopia: you get ever more difficult, more comprehensive, more incisively stated problems to solve. By great good fortune, we have progressively greater access to the comprehensive design principles of the Universe with which to solve these problems. It is undeniable that we humans have this local-Universe function. It is reasonable to assume that is why we are here. It seemed to be a very good working assumption. It has served me well for the last half century.

In order to avoid rousing the fears and consequent active opposition of the powerful financial, religious, and political interests who might foresee in my artifacts revolution the obsolescence of their own profitable products or services, I deliberately designed far into the future. I confined my complex of omniintercomplementary artifact-designing to function only within a socioeconomic era so many technological-evolutional stages further ahead in the future as to render the only-synergetically-effective interfunctioning of the many seemingly uninterrelated artifacts entirely unanticipatable by the overspecialized viewpoints of the pre-1929 economic world's most astute masters or of their most farsighted advisers. For instance, who in 1927 could foresee the intercomplementation of my various inventions: the cartographic projection of the world; my one-piece stampable bathrooms; my synergetic geometry; or my air-deliverable, mast-suspended dwelling machines?

Careful study of my anticipatory strategy for avoiding the incumbent world power structure's opposition to my only-in-distant-time-integratable artifact revolution placed the era for such safely immunized practical realization of a sustainable high standard of living for all Earthian humans as beginning sometime between 1980 and 1990. The latter was the date by which my design science's comprehensive inventory of artifact specifying and schematic inventing could be completed as initiated, organized, and maintained only by the single individual, demonstrating what happens when one invents always and only with total humanity in mind, along with total physical resources of planet Earth and the total cumulative technoscientific know-how, know-what, and know-when of all human history.

In production management there is a fundamental order of ''lags''—i.e., invention-to-use gestation periods—which relates directly to velocities operative in the respective phenomena considered: the slower the action, the longer the lag. In electromagnetics, where the velocity is 700 million miles per hour, there is a lag of only a few months between invention and industrial use. In the astro-aeronautical arts, where the velocities range from a few hundred to a few thousand miles per hour, there is an average five-year lag between invention and practical industrial production use. In the automobile arts, where the average velocity is only 60 miles an hour, there is a ten-year lag between invention and industrial use. In the skyscraper-building arts, where the highest velocity of motion is that of the completed structure's rate of heat- and cold-caused expansion and contraction and yieldings to hurricanes, which is measured in mere inches and fractions of an inch, the invention-to-industrial-production lag is one quarter of a century. In the production and operation technology of single-family dwellings, which are relatively immobile, there is a fifty-year lag between invention and use.

This anticipated fifty-year lag in the gestation of single-family livingry[1] technology happened to coincide neatly with the fifty-year minimum immunization period I adopted in 1927 to avoid the incumbent power structure's anticipatory opposition to an artifact revolution. (See the charts in my first book, published in 1938, *Nine Chains to the Moon.*)

When I say *artifact,* I mean any participation using the principles of nature to reassociate these principles for a specific purpose. Nature, for example, does this: she takes her own rocks apart. Nature is ceaselessly transforming.

The methodology of my artifact revolution is quite simple. Taking nature's cue, I determined that I must commit myself to solving problems by artifacts: what I call reforming the environment rather than trying to reform human behaviors. The function of what I call design science is to solve problems by introducing into the environment new artifacts, the availability of which will induce their spontaneous employment by humans and thus, coincidentally, cause humans to abandon their previous problem-producing behaviors and devices. For example, when humans have a vital need to cross the roaring rapids of a river, as a design scientist I would design them a bridge, causing them, I am sure, to abandon spontaneously and forever the risking of their lives by trying to swim to the other shore.

Having committed myself to developing physical artifacts which would reform physical circumstances instead of trying to reform human

---

[1] My term, to differentiate it from weaponry (or "killingry").

customs and the socioeconomic-political system, I faced another problem: it was obvious that to be realized, these physical artifacts were going to require costly materials, skilled craftsmanship, energy, and all kinds of tools and workshops. Since I was penniless, the number-one question was "How can I undertake such a fifty-year, world-embracing commitment with little or no money?" The logical answer to that question evolved in the following manner.

First, I was deeply impressed with what my scientific training had taught me regarding ecology and the fact that a great deal of energy is required to produce and sustain biological life on our planet.

Second, it was apparent to me that human beings must have some very important function to perform on planet Earth and in Universe—as I have already described.

That function and the human organisms which perform it require much energy. The Sun is planet Earth's greatest source of energy. However, human beings, being mammals, cannot acquire this life-sustaining energy through sunbathing. Solar energy must be gotten indirectly.

The planet Earth's botanicals convert their random, disorderly, entropic radiation-receipts from the stars—primarily the Sun—and then angularly rearrange the divergent radiation into convergent pattern integrities with beautifully ordered, syntropic[2] atomic and molecular structures—i.e., the hydrocarbon molecules used by all the discrete botanical species respectively in their unique, evolutionarily ordered growths.

These botanically harvested, evolutionary-structuring hydrocarbons and their constituent atoms—together with all those atoms' unique behavioral characteristics—are then superficially consumed and multiplied by the vast variety of hydrocarbon-hungry, mobile zoologicals. Sum-totally, the intershuttling of the mobile zoologicals—busy as bees in their travels—inadvertently but effectively cross-fertilizes the remote-from-one-another, rooted botanicals.

The complex, comprehensive, interregenerative system thus produced we speak of as ecology.

Generalized, Scientifically Verifiable Principles

The Earth's ecology is in such exquisite balance—with all its elements so interconnected and interdependent—that it appears seamless.

---

[2] My term for going to states of higher order; the opposite of entropic.

Observing this careful balance, the human mind gathers experiential evidence to intuitively project the same orderliness and connectedness onto Universe, surmising that the terrestrial order comprises a subset of a Universe that operates on pure principle.

This Universe of pure principle is so exquisite and absolute as to be perceived by the brain-coordinated human senses as constituted of altogether solid objects and organisms, even though no event or system touches any other event or system, with the atomic nucleus as proportionally remote from its electrons as the Earth is from the Sun.

One such pristinely generalized principle is that of interference. Conventional academic (Euclidean and post-Euclidean) geometry misassumes that a plurality of lines—more than one—may pass through the same point at the same time.

But a line is a trajectory of an energy event. If two events converge in the same location at the same time, an interference occurs, resulting in either a reflection, a refraction, a smashup, or a conjoined line of travel.

Experiments employing billion-dollar atom-smashers have demonstrated this fact. If lines could transit the same point at the same time, light rays would pass through objects and would not reflect back from an object to enter our eyes, and there would be no vision. Our vision requires interference between light and the surfaces of objects—more properly described as event complexes, since surfaces are always high-frequency event fields or grids. Because of this unfailing interference between lines of light and surface lines, the light rays bounce back to enter our optical system; that information is then quickly transmitted to the brain.

Another manifest of the same principle is the kinetic barrier produced by the invisible high-frequency motion of an airplane's revolving propeller blades. Even so, a machine gun can be coordinatedly timed to shoot so that its high-speed bullets pass through this kinetic barrier. On the other hand, the relatively slow speed of human arm motion makes it impossible to insert a hand between the revolving propeller blades and then withdraw it in time to avoid injury. A human can throw a baseball at a revolving airplane propeller and it will inevitably bounce back. It may be possible to throw a baseball fast enough to have it pass through an airplane's revolving propeller blades but probably not quite fast enough to avoid having one of the blades hit the baseball a glancing (refracting) blow, thus angularly diverting its path—a foul ball, in other words.

Instead of a machine gun whose firing is synchronized to shoot through the openings between the successive cycles of an airplane propeller, we can use a baseball-pitching machine and a propeller to illustrate the principle of relative frequency. Baseball-pitching machines are used in batting practice by baseball teams. Pitched-ball speeds can be accurately controlled with such a device.

As we stand and face a revolving airplane propeller, we recognize that the top ends of the blades move rightward and the bottom ends leftward. The farther out from the propeller's hub we look, the more space intervenes between the blades and the faster is its rightward or leftward motion—with the center of the hub turning ever more slowly rightward or leftward and with a theoretical absolute center hub point that is moving neither right nor left. Such a motionless axis can be optically and physically proven to exist only four-dimensionally by the dynamic vector equilibrium model—the ''jitterbug'' (my geometrical model, not the dance). With the jitterbug humans can for the first time demonstrate omnidirectional wave pulsation, as we will see later in this book.

Further, we can recognize that the extremities of the propeller blades are first sucking and then thrusting a volume of air through the blades and that the farther outward from the hub, the more powerful and high-speed the motion of the sucked and thrust-through air column.

If we aim our propeller-synchronized baseball-pitching machine's high-frequency-operating trajectory successively outward from the hub of the propeller, the pitched balls will encounter successively greater air-column suction and blowing forces. If we move the baseball-pitching machine somewhat to the side and aim it to pitch the baseball slantwise through the propellers' suck-thrust air column, the baseball's line of trajectory will be progressively deflected as it passes otherwise untouched through the revolving propeller. The more slantwise we shoot the baseball through the propeller, the lower the frequency of impacting and the greater the angle of deflection.

If we now take a sheet of parallel-ruled paper and draw a line with a straightedge laid perpendicular to the uniformly spaced parallel lines, we will have a diagram of the baseballs being pitched perpendicularly through the propeller. If we slant our straightedge and draw successively more slantwise lines, we find the distances between the parallel lines crossing those slantwise lines, to be ever greater.

Scientists make X-ray diffraction gratings consisting of tiny parallel grooves scored into the surface of a sheet of glass. The sides of the grooves are tilted at various angles not only to discover the interference variations resulting from such angle-produced, progressive widenings of the intervals but also to find the exact wavelengths and frequencies of the radiation examined.

Each groove in a diffraction grating is like a prism. The cross section of a prism of glass is a triangle. The sets of lines evenly parallel to the baseline of the triangle become progressively shorter as they occur ever nearer to the triangle's apex opposite the base. When a column of light passes through a prism of glass, the rays nearest to the bottom of the triangle pass through a greater number of atomic "electron-around-nucleus" propellerlike systems. The wider the glass, the greater the angular deflection of the radiation.

The many local propellerlike atomic-energy events of the glass prism structure operate like the parallel-arranged sets of pins in a pinball machine. This analogy holds true for annular distance variations of paper and straightedge intervals and for the X-ray diffraction grating interference with variations in "propeller blade" frequency. We can then comprehend how it happens that the trajectories of photons of light passing through the thickest part of the triangular glass prism get bent toward wider angles than those passing through thinner parts; and so we see why when the wavelength is most retarded it appears red and then, as the angles become narrower and the wavelength less retarded, orange, yellow, green, blue, and violet.

We can now understand, for the first time, why the Sun's rays passing at a low angle through the most atmosphere at dawn or twilight are reddish, while those passing more perpendicularly through the least atmosphere are bluish. And understanding this principle, we see the differentiating process of the colors of the rainbow.

The principle of relative frequency of interference and angularly diverted courses of travel is operative when light transits the myriad of regularly interspersed, locally repetitive, atomic-energy events that comprise our seemingly solid eyeglasses, which act like the alleyways between local pins in a cosmic pinball machine. Light moving at 186,000 miles per second can penetrate bumpingly from side to side through the "pin alleyways" in eyeglasses with only small angular and frequency changes of course, which by optical design can be refractively reangled to produce corrected eyesight. It is thus in pure principle that we can see through seemingly solid objects.

We must dispel our notion of solidity. Scientific experiment has demonstrated irrefutably that what continue to appear as solid objects to us are composed of atoms which are as relatively distant from each other as are planets in our solar system. Interferences and diverted courses of travel give a clearer picture of what happens when solid objects encounter each other. To reiterate, lines cannot pass through the same point at the same time.

Relative frequency is another way of viewing relative size. Take, for example, a cigar-shaped steel object 6 feet long and 4 inches in diameter; the object thus has a length-to-thickness (slenderness) ratio of 18–1 and weighs proportionally so much that it sinks swiftly in water. If we reduce this object to a length of 1 inch but maintain the 18–1 slenderness ratio, we will have a steel needle with a shaft diameter of $\frac{1}{18}$ inch, an object that floats on the water. The surface-to-weight ratio has changed dramatically.

Mathematically, this situation is expressed by the fundamental consideration that doubling the linear measurement of a symmetrical polyhedron is an increase at an arithmetical rate—i.e., at the first-power rate (n)—while the surface area increases at a second-power rate (n^2) and the volume increases at a third-power rate (n^3). In other words, as an object measured linearly increases in size at a ratio of 1 to 2, its surface area increases at a ratio of 1 to 4 and its volume increases at a ratio of 1 to 8. Thus, when an object's length is doubled, its surface area is quadrupled and its volume is octupled.

Physical behaviors of Universe vary greatly as size and frequency vary, though the principles are constant and eternal. Guy Murchie, in his 1981 book *The Seven Mysteries of Life,* points out that a mouse can fall unharmed from an airplane at great height, its skin-surface-to-weight ratio being that of a man with an opened parachute. Because of the same principle, an elephant falling from an airplane at great height would splatter on landing like a june bug on a speeding automobile windshield. This principle of greatly varying life behavior dependent on *relative size* was discovered by Galileo and named by him "similitude."

Long ago, clipper ship owners discovered that doubling the length of a ship increased its payload eightfold but the amount of ship surface to be constructed and driven through the sea only fourfold, thus halving the amount of energy (and expense) per pound of payload necessary to drive the ship through the sea.

This principle of similitude persuaded these capitalists to venture their wealth in building ever-longer, ergo ever-larger, ships of the sea, which in turn ultimately led to their controlling and monopolizing the planet's lines of supply.

The method for producing ever-larger ships was first to build the keel, ribs, and skin of the hull in dry dock and then, after launch, move the hull from one outfitting dock to another. Local acquisition of vital parts for the sailing ships was followed by an around-the-world series of acquisitions: stronger masts when docked in a country with superior knot-free wood, rope when docked in locales known for the strongest hemp, and so on. This moving production line became the prototype for all present-day mass production—that is, moving assembly lines.

The principle of similitude (or relative size advantage) also motivated bankers to amass capital by employing the funds banked with them by unwitting depositors to achieve the vast magnitude of resources required to take advantage of similitude—doubling the length of their ships, thereby fourfolding their profit. The person who invented and produced a keel-and-rib rowing boat, though possessed of the knowledge of the principles involved, never had sufficient capital to build the big ships. Then as now, profits derived from the ingenuity of inventors are usually realized only by the owners of mass capital.

Artifacts: Application of Pure Principle

This principle of relative size advantage is not popularly understood. Indeed, despite the economic importance of the principle of similitude, it has yet to be incorporated into university engineering curricula.

In 1954 I patented the geodesic dome, a new structural system that solved centuries-old architectural problems of enclosing space and spanning distance. The "omnitriangulated" structural principle of the geodesic dome was described by the American Institute of Architects, in their Gold Medal citation, as "the strongest, lightest, and most efficient means of enclosing space yet devised by man."[3] It is the only structure we know of that gets stronger as it gets larger and has no limit to its span.

When we double the diameter of a geodesic dome, we increase the volume by a factor of 8 and the surface by a factor of 4. This means we enclose eight times as many molecules of atmosphere with only a fourfold increase in the enclosing skin through which that atmosphere can gain or lose heat. Doubling dome size doubles the thermal efficiency of domes while halving the amount of enclosure that needs to be built.

The economic importance of these mathematically derived principles remains unknown and undiscussed in the academic world. I felt that many of the world's most serious problems were rooted in the ignorance of how to apply these principles to solving problems in the real world.

With this working assumption regarding the eternal reliability and

---

[3] R. G. Wilson, *The AIA Gold Medal* (New York: McGraw-Hill, 1984), p. 210.

absolute reality of pure principle, I intuited in 1927 that it might be possible for me to commit my energies to the realization of artifacts of a physical environment whose performance per unit of invested resources would be so comprehensively improved that it would free humans from the competitive struggle to exist and thus encourage humanity's spontaneous cooperation to achieve and sustain mutual physical success for all—that is, an unprecedentedly high standard of living for everyone on the planet.

My intuition seemed to describe an evolution that is intent upon developing humans to the point where they can achieve total physical success. At that point, humans could become preponderantly preoccupied with—or, more correctly, could act upon—the exclusively mind-solvable problems attendant upon supporting the integrity of eternally regenerative Universe.

In view of all the foregoing considerations of principles and assumptions of cosmic purposes, I recognized that it might well be that as a mere individual, I would need no planetary socioeconomic authority's approval for my undertaking and that if I conducted it effectively, my work would be economically sustained in entirely unexpected, unsought-after ways.

I observed that in nature's own economics, that of ecology, the grass was not obliged to pay the clouds for rain. Regeneration, being comprehensive and interdependent, neither gained nor lost energy and could only grow sum-totally in the realized wealth of ever-greater know-how and wisdom.

This observation answered my number-one question. It seemed to me that I was clearly informed on how to proceed. If and when I was doing first what first needed to be done, working out the most effective strategies in pure principle, I would be able to carry on successfully. If I was not doing things in proper order or doing irrelevant things, I would be unable to carry on. If I was not getting along, I would change course and look for a way to return to smooth sailing.

With the backing of "great intellectual integrity," I would require no other support. My support would be in exercising the operation of the comprehensive set of all omniinteraccommodative generalized principles of eternally regenerative Universe. My support might show up as money or materials or tools or workshops or whatever else might be needed.

Only time and sustained commitment would tell me whether my principal working assumptions were correct. I posited, for example, that humanity was entering an unprecedented state of comprehension of principles and mental competence adequate to the epochal inception of conscious, spontaneous, voluntary realization of magnificently essential, new-to-Earthian-humans, functioning in Universe. This new stage of human evolution was no longer automatic, but a matter of conscious will.

Looking for confirmation of my many working assumptions, I returned to terrestrial ecology. I noted that vegetation had to be rooted in order to (1) expose enormous amounts of foliage and not be knocked down by the great winds and (2) draw water from the ground through roots with which to structure itself as well as to return waters to the sky regeneratively to structure and energize all terrestrial life.

Because vegetation is rooted, it is prevented from reaching other vegetation for purposes of procreation. For this cross-fertilizing purpose, nature designed the mobile zoological hosts of subsurface-boring, surface-crawling and -walking, and air-flying creatures to traffic back and forth among the rooted botanicals. Nature chromosomically programmed the zoologicals to go after honey or other metabolic rewards and only incidentally to cross-pollinate the botanicals. Nature did not say to the honeybee, "I want you to go out and cross-pollinate." Nature, through DNA-RNA coding and chromosome-level programming, said to the honeybee, "Go after honey," knowing that the honeybee must inadvertently cross-pollinate with its bumbling tail, inherently facilitated by the purposefully designed proximities of the flowers' vital-organ arrangements.

Human beings have been designed to be born naked and helpless. They are given comprehensive regenerative equipment but, having no experience, are absolutely ignorant. They become hungry, thirsty, and curious, and in usual course have a procreative urge, all of which "drives" or forces cause them to take initiative and thus learn. This learning takes place gradually, often at great expense and exasperation, and only by trial and error. Eventually humans learned how to domesticate animals and vegetation.

Let us consider the case of a human being who is a milk cow breeder and herder. He has ten children, all of whom need milk. He has cows enough to take care of not only all his children but also those of a hundred other families. However, his ten children also all need shoes. There exists a man in the same tribe who has developed the ability to make shoes from cowhide. The shoemaker can make many more shoes than he and his family of ten milk-thirsty children can wear. The shoemaker wants milk for his children. The cowherd comes to the shoemaker. They realize that they cannot cut up the cow and still milk it.

They talk over their needs as well as their experience in producing their respective products. They agree that the cowhides for the shoemaking become available from cattle that are not being used for milking and therefore do not enter into their particular trading problem. They agree that it takes very much longer to produce a milk cow than it does to make a pair of shoes.

To accommodate such exchanges of disparate goods, humans invented money. Money consisted of tokens made of substances of no intrinsic value—such as white pebbles or beads—which all of the tribe recognized and accepted as representing easily counted tokens for purchasing capability and as calculating devices representing the holders' input into the community wealth. This wealth was realistically accounted as being the capability to support, protect, and accommodate forward days of various numbers of human lives.

Money "beads" realistically represented the accountable hours and days of human production or work invested in the respective exchange items and services. The tokens could be set aside until needed.

In ascertaining nature's economic principles, I next recognized that the principle of laying the credit tokens to one side demonstrated how nature often operates at 90° (that is, sidewise). In railroad operations this is called shunting. It allows society to sort out its resources and to selectively time their interaction. The shunting can be accomplished by veering off into a sidetrack, or it can be accomplished by deliberate right-angle setting aside into a local holding pattern.

The honeybee inadvertently bumbles off its cross-pollinating perpendicularly to its chromosomically programmed line of action. Ecology is comprehensively interregenerative at 90° to the "in-front-of-your-nose" line of attraction. For instance, the honeybee aims at the nectar-sack and inadvertently knocks off the pollen sideways—at 90° to the direct line of honey approach.

Humanity likewise can be seen to be chromosomically programmed to act like honey-money bees—continually buzzing in and out of attractive situations in search of honey-money with which to support self and family. Humanity then, inadvertently, through fear-supported government war budgets, produces the high-efficiency technological production facilities that are reserved for weaponry and government-sanctioned murder but, fortunately, in due course, are used for the right life-supporting reasons. For example, electric refrigeration, first used on battleships, is adopted a generation later for use in the domestic environment on dry land. Here again is the principle of similitude at work—adequate capital made available only for life-or-death defense armaments. Human beings, while apparently working at cross-purposes, do the right things for the wrong reasons—inadvertently—in a precessed (sideways) manner. Of course, acting with conscious direction is the next stage of human evolution. I call this discipline anticipatory design science.

The Bible speaks of the postwar conversion of swords into plowshares. If the metallic plowshares had been produced in the first place, sufficient food production for everybody would have been possible. Lack of food and other life support brought about the fighting to begin with. Those suggesting production of metal plowshares before the war were always given the brush-off by tribal or state leaders and told, "Metal plowshares are far too expensive. We shall make do with wooden ones." The peacetime economy was differentiated from the state "on a wartime footing." In the long view, however, heroic expenditures for basic life-support needs make good economic sense.

It became apparent to me that in its primitive stages nature attained its energetic regeneration only inadvertently—only by its 90° "side effects." Nature employs the 90° effects comprehensively in its magnificent regenerative design manifest—the right-angle principle of which is called precession.

What appears on first viewing to be linear motion is seen in the greater view as the cyclical motion of regeneration.

The precessional effect of the Sun's motion on its gravitationally retained planet Earth makes the Earth orbit the Sun in a path of 90° to the line of gravitational interattraction; so, too, does the electron orbit around the atomic nucleus, a manifest of pure principle.

As we will detail later, there are six positive and six negative degrees of freedom in Universe in respect to which all structural systems in Universe must abide. Every healthy and active child quickly discovers five of them, as more fully described in *Synergetics* (1975): (1) axial rotation, (2) orbital rotation, (3) expansion-contraction, (4) torque (twist), and (5) "inside-outing." The sixth, precession, is also experienced by the child, most clearly in the realm of toys: the child's top, during its fast axial spinning, also leans away from its axis, revolving in this half-fallen attitude, without any witnessable tendency to fall further. This precessional behavior is also manifest by a toy gyroscope, which can spin on the end of a pencil while leaning precariously. Not only do children find nothing in their other experiences to explain these "peculiar" and "exceptional" behaviors, but neither do all the professors of science. Because scientists have had physical experiences that defied their capability to explain in strictly sensorial terms but could be reconciled through the use of mathematical formulae employing quantum mechanics, science in general determined that only mathematical formulae should be used by pure scientists and that models were dangerously illusory.

I have always found models quite useful in illustrating apparently complex phenomena in nature. For instance, I have found the models of synergetics, my system of geometry, quite capable of illustrating such basic principles as quantum mechanics, fourth-dimensional forms, and complex motions and phase transformations.

From 1938 to 1940 I was on *Fortune's* staff as the science and engineering consultant. In late 1939 I prepared an article on the Sperry Gyroscope Company which appeared in May 1940. Mr. Bassett, vice president of Sperry's engineering department, pointed out that the American naval and air forces used many gyroscopes for both directional compasses and directional control mechanisms. Although told by the president of Sperry that precession, the heart of the story, could be explained only in terms of the mathematics of quantum mechanics, I presented a two-page explanation of precession in terms of human senses rather than mathematically abstruse formulae, as I have done from the lecture podium many times since.

The fact that precession occasioned science to adopt only mathematical formulations for all its pronouncements makes clear that precession's sensorial explicability should also occasion science's return to sensorial procedures. In *Synergetics* I set about to do just that. Science has not yet yielded to models, but it will, returning mathematics to a more comfortable relationship with the everyday world.

The only explanation of precession thus far written in realistic—that is, sensorial—terms is the article on Sperry that I wrote for *Fortune*. Here follows an even more concise sensorial and modelable explanation of precession.

There are two kinds of physical acceleration, linear and angular. The field athlete known as a hammer thrower uses *angular acceleration* to accumulate the energy he exerts to build momentum in his steel sphere "hammer." Hammer throwers use their muscles to accelerate the hammer. They use the muscles of their arms and their hands to tightly grip the triangular handles attached to the end of the steel rod that is connected at its other end to the heavy steel ball called the hammer.

Olympic hammer throwers must stay within a circle that is clearly marked on the ground and is just large enough to allow them to use their leg and back muscles to rotate their bodies while tightly gripping the handles of the hammer. Hammer throwers thus angularly accelerate the ball as they rotate their bodies. After the permitted amount of rotation, during which the hammer and its control rod are angled at 90° to the line of desired travel, the thrower releases the hammer.

The hammer thrower's rotating motion elevates the hammer from the ground, swinging it around at an ever-greater elevation and at an ever-increasing circumferential speed until the hammer is finally rotating at the athlete's shoulder height. The more muscle energy the athlete invests in the acceleration, the farther will the released steel hammer travel before landing. When the hammer thrower lets go, the hammer travels away tangentially at 90° to the circle of its acceleration. Thereafter, until landing, the hammer is operating in linear momentum.

A tennis player angularly accelerating his tennis racket around his own center of gravity hits the tennis ball, which is then linearly accelerated toward the net. The bullet in a gun is linearly accelerated. The molecules of water in a garden hose are linearly accelerated.

Linear acceleration does not accumulate momentum and is progressively expended. A spaceship rocket is linearly accelerated, as successive multistage explosive linear accelerations enable it to attain exit velocity and escape from the density and friction of the Earth's atmosphere. Once the rocket is in orbit, the gravitational pull of the Earth and other celestial bodies is only *radial* or *angular* and, like the arms and steel rod of the hammer thrower, has no axially fricative, acceleration-retarding, or energy-expending effects on the initially linearly accelerated body.

Celestial bodies always travel orbits in a direction at 90° to gravity's tensional pull on the orbit. The orbited-around body is the gravitational master body.

Going back to our first example, we recognize that what the hammer throwers muscularly contend with are (1) gravity's constant downward pull both during the acceleration and after release of the hammer and (2) air resistance to the hammer athlete's angular accelerating as well as to the hammer's released-in-flight, linear, through-the-air travel. As a consequence, the pattern of overall travel of the released hammer on a windless day is that of a quarter ellipse in a vertical plane, with the hammer constantly slowing in its horizontal travel and finally decelerating into exactly vertical travel toward Earth.

Abruptly leaving the hammer thrower, we will now consider a peashooting device driven by compressed air. This device causes the linear acceleration of unit-radius plastic "peas" blown out through a tube whose diameter is just an invisible increment greater than that of the plastic peas passing through it. Again assuming a windless or draft-free environment, we witness the peashooting machine aimed due north and parallel to the ground. Gravity gradually pulls the shot-forth peas' trajectory Earthward all along its northward route of forward travel. Each blown pea travels along a path describing a quarter ellipse in a vertical plane and ending in a vertical descent to Earth. If we stand close to the plastic peashooter's nozzle and insert our finger into one side of the pea trajectory near the mouth of the shooting tube, we find that we can deflect the exiting peas' trajectories in various ways.

Putting one's finger exactly in front of the tube opening will completely arrest the peas' linear acceleration; now accelerated only by gravity, the peas will plummet perpendicularly to the ground. We can also move our finger in from one side of the trajectory and very gently touch the bottom of a train of accelerated peas, causing them to rise very slightly. Now, resting our hands on a slidable side table, we extend our index finger beyond the table's edge in a fixed position touching the right side of the trajectory of exiting peas. Thus, the peas' horizontal trajectory is deflected leftward, to the north-northwest while also, as always, being pulled groundward by gravity. So long as our finger remains in this fixed position, it will continue deflecting the horizontal path of the linearly accelerated peas, each of which will keep on describing the same one-quarter ellipse in a vertical plane aimed north-northwest.

In Fig. 1.1 we see the train of uniform-diameter plastic peas being blown out of the peashooter. We see the human finger intervening delicately into the train and deflecting the train. We note that no pea has a memory that directs it to resume its earlier direction of travel.

What we learn from the foregoing is that after being deflected, a pea (or any other body in acceleration) does not resume its earlier course. It has no memory of its earlier travel pattern. It continues to be affected only by (1) the initial acceleration, (2) the friction and density of resistance of the medium penetrated (in this instance, the air), and (3) the

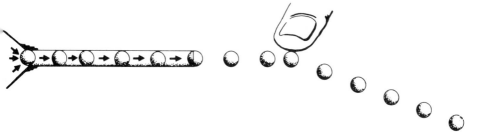

FIG. 1.1 *Peashooter and downward deflected pea trajectory.*

last angular redirection of its trajectory, such as a crosswind's gust or a deflecting contact with a finger.

If, instead of deflecting northwestward the initially northward accelerated peas, I were to bring my finger down exactly vertically 1/32 of an inch on top of the peas' northward path and keep my finger exactly in this position, I would deflect the trajectory mildly downward.

Now we return to observing the hammer-throwing athletes. In Fig. 1.2 we look at the hammer thrower at peak angular acceleration of the hammer. We note—and this is very important—that he releases the hammer when it is tangent to the circle of his gyration, so that it travels in the direction he wants it to travel when he releases his grip. Unlike the javelin thrower and the shot-putter, he does not release his thrown device at 180° (in the direction in front of him). The hammer thrower (like the discus thrower, tennis player, and baseball batter) ends his angular acceleration at 90° to the desired line of travel of his hammer— that is, when it is 90° short of the direction of realized acceleration. Linear acceleration terminates at a point that is in a direction exactly 180° away from the accelerator.

Let us assume that the formal Olympic Games–determined direction in which the hammer is to be let go is true north. Thus, as angular-accelerators, the athletes are going to let go of the hammer when they

FIG. 1.2 *Hammer thrower.*

are facing true east. We will make a simple mechanical model of this event. The model of the hammer thrower will be a vertical ½-inch-round steel shaft 6 inches high; the hammer will be represented by a steel ball ½ inch in diameter; the thrower's arms and hands and the steel rod leading out to the hammer will be represented by a round steel shaft ⅛ inch in diameter.

We will now take a ½-inch-thick circular steel ring 8 inches in inside diameter and 9 inches in outside diameter.

This ring has two short cylindrical housings mounted on the inner surface at both the top and bottom. These housings contain compressed-air turbines and tapered roller bearings to drive and align the "thrower." The tapered roller bearings in these housings now receive the top and bottom, respectively, of the ½-inch-diameter, 6-inch-high vertical steel shaft representing the hammer thrower.

The thrower's body, represented by the vertical shaft, constitutes the axis Y and the ⅛-inch-diameter steel rod representing the thrower's arms and hands grasping the triangulated handles attached to the steel ball hammer constitutes the X axis in Figs. 1.3–1.7.

The air turbines are driven by compressed air supplied through ducts in the hollow steel A ring (labeled A in Figs. 1.5–1.7). The compressed air is continuously ducted through hollow tubular shaft bearings at axes X and Z and through the hollow B ring, the hollow half-round C ring, and the base of the whole three-axes-of-circular-freedoms apparatus shown in Fig. 1.7.

In Fig. 1.3, A is a bird's-eye view of B, which is axis Y, the model of our hammer thrower with the air turbine in operation and the hammer whirring around axis X. We have a ¹/₁₀,₀₀₀-second view of our hammer thrower at a moment when his hammer, H, is extended toward you and me, the viewers.

In Figs. 1.4–1.6, we have the same ¹/₁₀,₀₀₀-second flash glimpse of axis Y, with our hammer thrower revolving at so high a speed that his ball becomes in effect a flywheel, as seen in Fig. 1.7.

Our human finger now touches the top of revolving hammer H as it passes in front of us. This top touch deflects the hammer's line of travel downward and to the right. This deflection forces the thrower's head and his Y axis top to also rotate downward and to the right, while his feet and legs rotate up and left; this rotation is accommodated by the rotatability of axis X (see Fig. 1.7).

Using a complete wheel to reduce directional stresses in the apparatus, we learn that if we touch the top of the spinning flywheel at point T in Fig. 1.7, it will cause the wheel and the top of its axis Y to rotate

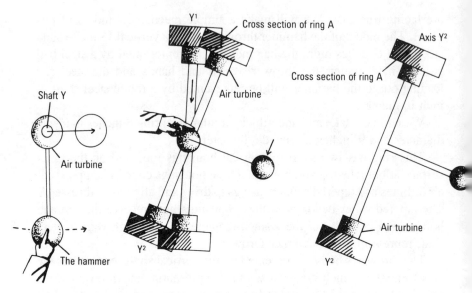

FIG. 1.3 *Mechanical model of hammer thrower.*

around axis Z. If, instead, we try to pull the top of axis Y left toward the left-hand edge of axis Z, we will witness the top of axis Y rotating right more or less around axis Z, toward us. It is this natural yielding in a direction at 90° instead of the expected 180° to the direction of force that has made the gyroscope so perversely incomprehensible to our senses-coordinating brains.

This yielding-at-90° phenomenon is known as precession. Its inherent incomprehensibility persuaded physicists to assume that it could, in the end, only be explained and manipulatingly coped with through the mathematical formulae of calculus and quantum mechanics. Because there existed an area of physical experience that seemingly could not be explained in sensorial terms, academic science concluded that the phys-

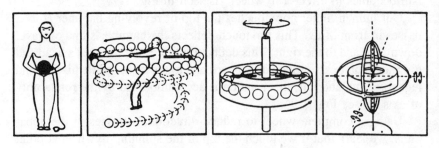

FIG. 1.4 *Schematic drawing of hammer thrower.*

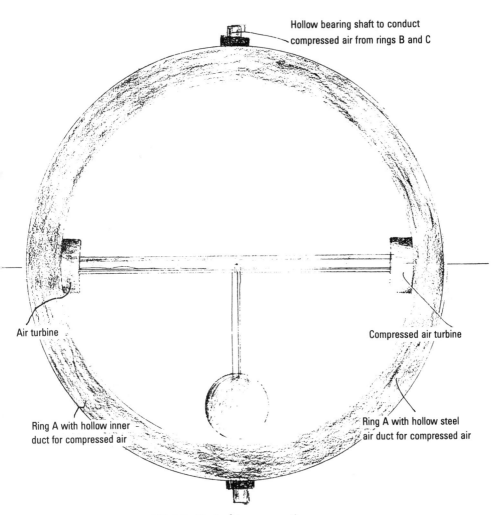

FIG. 1.5 *Air turbines operating.*

a. Hollow bearing shaft to conduct compression air from rings A, B, and C
b. Compression air turbine
c. Hollow steel air duct
d. Ring A with hollow inner duct for compression air
e. Air turbine

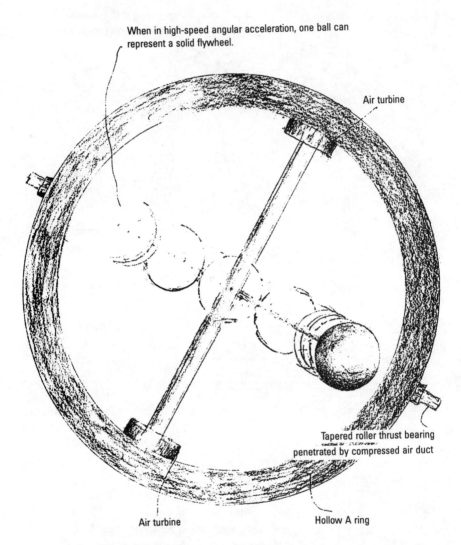

When in high-speed angular acceleration, one ball can represent a solid flywheel.

Air turbine

Tapered roller thrust bearing penetrated by compressed air duct

Air turbine

Hollow A ring

FIG. 1.6 *Model illustrating high-speed angular acceleration.*

a. When in high-speed angular acceleration, one ball can represent a solid flywheel.
b. Air turbine.
c. Tapered roller thrust bearing, penetrated by compressed air duct
d. Hollow *A* ring
e. Air turbine

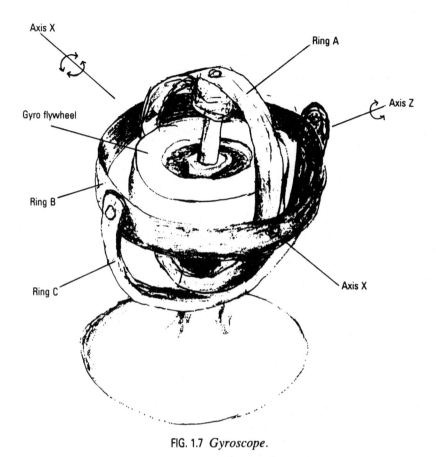

Axis X

Ring A

Axis Z

Gyro flywheel

Ring B

Ring C

Axis X

FIG. 1.7 *Gyroscope.*

ical world's behavior could be comprehensively coped with only (without exception) through equations and calculus.

Sensorially comprehendible precession makes lucid much of wave theory and electromagnetics (see my 1982 book, *Tetrascroll*).

In the social sphere, precession accounts for humans not yielding at 180° but yielding at 90°, and thus the orbiting of the less-powerful around the more-powerful in the various classes considered.

Doing the right things for the wrong reasons is typical of humanity. Precession—not conscious planning—provides a productive outcome for misguided political and military campaigns. Nature's long-term design intervenes to circumvent the shortsightedness of human individuals, corporations, and nations competing for a share of the economic pie.

Fundamentally, political economists misassume an inadequacy of life support to exist on our planet. Humanity therefore competes militarily to see which political system—socialism or capitalism (less exactly known as free enterprise)—is fittest to survive. In slavish observance of this misassumption, humans devote their most costly efforts and resources to "killingry"—a vast arsenal of weapons skillfully designed to kill ever more people at ever-greater distances in ever-shorter periods of time while employing ever-fewer pounds of material, ergs of energy, and seconds of time per killing.

Initially unforeseen, the mass-production technology acquired supposedly only for weapons making has been converted after each war to powerful and practical use. Cosmic evolution has put humans to work developing, unconsciously, the technology to produce ever more effective results in ever-quicker ways at ever-greater ranges of effectiveness with ever-fewer pounds of material, ergs of energy, and seconds of time per accomplished function, thus inadvertently acquiring the technological capability to do what politics could never do—that is, to produce so much high-standard life support with so little material and energy investment as now, for the first time in history, to be able to sustain all humanity at ever-higher standards of living than any have ever before experienced or dreamed of.

Because, as with all children, I had been born a deliberate comprehensivist and because I had not had that innate interest stifled and had grown to be a deliberately self-cultivated comprehender of invisible as well as visible reality in an age of specialization, I clearly saw, and broadly announced, that all the foregoing was true and feasible.

I also realized that our newly achieved evolutionary stage of technical ability to sustainingly support all humanity at an unprecedentedly higher standard of living was, or is, to be accomplished only by objectively and synergetically integrating the vast ranges of invisible reality's electrochemical and electromagnetic spectra. The realization demonstrated to me the exclusively mind-over-technology functioning of humans in Universe.

It became clear that only by good fortune did I happen to first stumble upon this emerging potential when I did. In 1927, I was a complete failure by society's standards of economic success. I was about to commit suicide on the shore of Lake Michigan when it dawned on me that potential success, not just for the individual but for all humanity, was implied in doing-more-with-less, invisible-reality technology.

I was convinced that humanity was graduating into a new era of consciously direct—rather than inadvertent—evolution marked by the realization of its cosmic, intellect-conceived, design-science functioning advantage in Universe. Henceforth, and swiftly, we must progress to the stage of doing all the right things for all the right reasons instead of doing all the right things for all the wrong reasons, a by-product of precessional phenomena.

Einstein proclaimed that there are only two prime motivations for all human initiatives: fear and longing. Acquiring the costly technology for producing national-defense armaments alone is the politically assumed number-one mandate, a mandate based on national fear. Such a survivalist mentality inadvertently also produces life-supporting technology, but it takes a quarter of a century longer than it would if humanity first recognized the public longing to attain sustainable peace for all humanity and directly used that same high-technology production for livingry rather than for armaments.

I am convinced that nature uses different gestation rates for both biological and technological phenomena. I am also convinced of the infallibility of nature's revolutionary intertiming design of these different gestation rates.

In my personal strategy, I eschew all promotion for this reason. I have no desire to develop the "premature babies" of industrial technology. As a consequence, I have no literary agents, no lecture bureaus, no advertising or public relations people, no sales agents of any kind. Neither myself nor anyone on my staff is allowed to solicit supporting grants. I have no sales people who go out to sell me in order to fund an operating budget. I ask no one to listen to me or to look at what I have produced. I speak to people only when they ask me to do so. When, however, people ask me what it is they see that I have produced, I give them my very best explanations. These personal operating principles are based on a kind of self-sufficient mechanism that I have always appreciated in nature's designs—and some supply-side economists have admired in human institutions. These rules of thumb have carried me through many crises during the past fifty-five years.

My economic survival pattern was based on my fortunate assumption that nature would support me and my work but only if I eschewed all politics and worked entirely in artifact invention and development and only on behalf of all humanity.

In view of all the foregoing, I saw the work of Albert Einstein as that of an individual who seemed to have been uniquely inspired by a clear vision of nature's generalized principles. I found myself to be inspired by an awareness of the evolutionary significance of the human mind's winnowing out of those generalized principles and the synergetic consequence of the objective reintegration of the Universe of principles into a myriad of local in-Universe special-case-evolution-through-problem-solving technology.

The era of human exploration and operation in the 99.99 percent of reality nondirectly contactable by the human senses is coincident with Einstein's realization that evolutionary change is normal and that the normal speed of all electromagnetic radiation is 186,000 miles per second.

This view completely altered for humanity the concept, established by Isaac Newton, that the physical norm is the state of rest. In this view, the physical norm is changeless, and thus, change is to be avoided.

When Einstein's concepts were first introduced, Professor Percy Williams Bridgman of Harvard, the pioneer in cryogenics, sought to understand why Einstein had caught the whole world of science so far off physically comprehensible balance. Bridgman concluded that the difference between the viewpoints of conventional science and Einstein (and their consequently employed methodologies) was that in contrast to science's attempt to isolate experiments within "controlled conditions," Einstein was always comprehensively considerate of all the environmental conditions and events attendant upon the experiment.

Bridgman called Einstein's methodological concern with both comprehensive and incisively focused-upon information "operational procedures."

I was excited to learn from Dr. Bridgman in 1947 of Einstein's operational procedures, for without knowledge of Einstein's having done so, I had come to share similar concerns and had in 1927 spontaneously adopted similar comprehensive concerns in my own work.

Operational procedures eliminate all recourse to axioms—the "it-has-always-been" or "it-is-assumed-to-be" truisms commonly employed by much of our educational system, particularly in those areas of education that most people think of as having long ago been infallibly explained by mathematics, physics, engineering, semantics, geography, meteorology, and cosmology.

I am convinced that academic science's comprehensive, three-dimensional, perpendicular-parallel, nonintertransformative, coordinate mathematics of "framed" referencing of all physical experiences is so awkwardly alien to nature's four-dimensional, convergent-divergent, discretely tunable, coordinately constant system as to render present-day academic science's mathematics unnecessarily complex and understandably incomprehensible to the majority of clear-thinking youth. As such, present-day science's inscrutability prevents us, who are laboring under the political-religious axiom that a fundamental inadequacy of life support exists on planet Earth, from spontaneously apprehending what has transpired in the invisible reality and thereby comprehending why and how it is now technically feasible to take care of all humanity at a sustainable higher living standard than any humans have heretofore experienced.

On the other hand, I am confident that I have discovered nature's own coordinate system. This most economical and popularly comprehensible, mathematical, intercoordinate, formative, energy-matter intertransformative, and deformative system is definitively presented in the approximately thirteen hundred pages of *Synergetics* and *Synergetics 2*. These volumes enable an individual to comprehend design science effectively and adequately.

2 DISCOVERIES OF THE HUMAN MIND

I WROTE *SYNERGETICS* BECAUSE I was overwhelmed by the experimentally provable evidence of what we have come to call synergy—i.e., the behavior of whole systems unpredicted by the behavior of any parts of the system when considered only separately. Synergy is antithetical to our society's preoccupation with specialization. I felt there was no concept more prominently conducive to effective thinking about the lesson-learning significance of the history of all humans' experience than is-and-always-has-been-and-will-be *synergy*.

To elucidate for you, I shall describe how I differentiate the function of brain and mind, as I first did publicly as the Harvey Cushing Orator of the American Association of General Surgeons at their annual congress in Chicago in 1968.

This differentiation developed as one of the consequences of my lifelong quest to discover and identify the function of humans in Universe.

In comparing humans with all other living organisms, it became clear that all living organisms other than humans have some built-in, integral, organic equipment that gives them an advantage in some special physical environment—for instance, the little vine that grows only along the banks of the upper waters of the Amazon or the dog with very short legs and nose close to the ground, allowing it to follow a scent trail, and with sharp claws to open the holes to the hiding places of its quarry. Birds fly in the sky with their beautiful wings, but when they are not flying, these wings greatly impede the birds' walking, because they cannot be discarded when not in use.

It was clear to me that if nature had intended to have humans function as innate specialists, she would have provided them with, for instance, organically integral telescopic or microscopic eyes.

Also clear was the fact that humans are not unique in having brains. Many creatures have brains. Brains are always and only coordinating the information of the senses—sight, hearing, smell, taste, and touch. Our brains provide the only means by which we are aware of "otherness" and ergo aware of being alive in Universe. Brains are always coordinating the sensed information regarding each special-case experience: this smells this way, that sounds that way. Brains always and only deal with special-case data, packaging them systemically and storing them for later recall.

Despite claims to the contrary, no one has ever seen outside self. We see only in our brain's "control room," with its omnidirectional television system. What we see there has proven to be so reliable regarding our surroundings that we now misassume that we are looking outside, seeing it "over there."

In contradistinction to brain, human mind manifests from time to time the extraordinary capability of discovering relationships between special cases of the sort not evident from examining any of the special cases alone. Mind discovers interrelationships.

While there is an impressive list of the human mind's invisible interrelationship-discovering capabilities, there are twelve cases that stand out.

The first was demonstrated in a complex of historical scientific discoveries and measurements that began with Copernicus, Kepler, and Galileo and culminated in Isaac Newton's mathematical formulations of the laws governing the covarying, invisible interattractiveness of any two celestial bodies. This invisible interattractive force varies inversely as the second power of the arithmetically expressed distance intervening between the two bodies considered, while the relative interattractiveness of any two celestial bodies in respect to that existing between another pair of celestial bodies is always proportional to the multiplicative products of the respective pairs' respective masses.

The second of these twelve historically most extraordinary manifests of human mind's invisible interrelationship-discovering capability occurred when a human mind discovered the desirability and complex calculating capability inherent in the mathematical symbol for nothing— the cipher. That unknown something, the x of algebra, is a conceivable "something," but an unknown, unitarily specific "nothing" is quite inconceivably different from all the other unknown nothingnesses of Universe. You cannot eat "no sheep." You cannot think of, or feel hungry for, a specific "nothing."

Only the Polynesian navigators' offshore orientation needs necessitated the invention of trigonometry for locating terrestrial sea position by observed and calculated intertriangulation between the boat's position and any two other remote fixed objects, such as any two stars in the sky.

From time to time, being subject to being washed overboard by gale-driven seas, these naked Polynesian navigators found it necessary to keep track of the cumulative scores of their fingers-and-toes ten- and twenty-increment counting. They did this by fastening sets of rings round their wrists, ankles, and neck. Each ring represented already counted bundles of ten fingers, ten toes, or both. This inventive use of sliding rings to represent cumulative decimal increments I am sure led to the invention of the abacus—a formalized and more-convenient-to-use device in the form of a framed, bamboo-rod-mounted, ring-bead calculator. In the Polynesians' ingenious precursor to the abacus the counters are the anklets, bracelets, and necklaces which would not be lost in ocean storms.

Only the foregoing could account for the operational-method-enforced leftward positioning to symbolize a leftwardly moved bead or modular increment of ten. From such a model, it is reasonable to assume, arose the mind-invented set of Arabic numerals.

To represent an empty column necessitated the invention of the cipher. It symbolized a uniquely unified, precisely interpositioned, immensely useful nothing.

In the mists of antiquity, human mind conceived the need for, and the operating mechanics of, the digital calculator, but surely not all of its future possibilities.

The third most-extraordinary manifest of the human mind's discovery and mathematical formulation of invisible interrelationships occurred when, prior to the French political revolutionaries cutting off his head, Antoine-Laurent Lavoisier intuitively reasoned that the invisible nothingness known only as the mystical element air was ignited within the bell jar of Joseph Priestley's experimental isolation of phlogiston ("fire"). The experiment-produced substances weighed more than the substances originally placed under the bell jar prior to ignition. This experiment caused Lavoisier to assert that the air under the bell jar consisted of a plurality of entities each so fundamental as to be identified as chemical elements. This was an extraordinary conception: the differentiation of the undifferentiated nothingness into identifiable gases—each so unique as to rate as a chemical element. Lavoisier did his thinking in an era when all the thus-far-discovered elements were metals—tangible and substantive. Lavoisier named one of the gaseous elements oxygen, which he said had separated out from the other invisible gaseous elements and had combined with the weighed-in substances, wherefore he proclaimed combustion to be "oxidation." He went on to substantiate his argument by demonstrating that rust is oxygen combined with iron and that separating oxygen from mercuric oxide produces the liquid metal mercury.

The fourth most-extraordinary manifest of human mind's ability to discover invisible interrelationships of Universe occurred when Democritus conceived of atoms.

The fifth most-extraordinary manifest of human mind's ability to discover invisible interrelationships occurred when Hertz discovered electromagnetic waves.

The sixth most-important manifest of human mind's discovery of invisible cosmic interrelatedness occurred when the mathematical working of gyroscopic precession was discovered by Elmer Sperry.

Human beings, for at least three and a half million years on board our planet, have observed the seemingly fixed constellar patterns of the starry skies. In stark contrast to the fixed stars viewed from Earth as members of stable constellation groupings, humans also sometimes observed one, two, perhaps three or more starlike objects, often a little brighter than the other stars and with a little more vivid coloration.

Appearing first in one fixed-star constellation on one night and then reappearing in another constellation the next night, these bright, wandering objects were obviously like the Moon, traveling in respect to the thought-to-be fixed stars. These were the planets, and humans gave them the names of gods and began superstitiously identifying the significance of the planetary appearances with various experiences of their Earthbound lives.

In the fifteenth century the South German and North Italian scientists acquired calculating capabilities made possible by the cipher and consequent positioning of numbers.

In Poland, Copernicus discovered mathematically that the Sun was not encircling the Earth but just the reverse. The Earth was in fact one of the planets orbiting the Sun.

Then, Kepler made very accurate observations of all planetary behavior and characteristics. The planets appeared to constitute a very disorderly team. Kepler plotted the position of each planet at the beginning and end of a twenty-one-day period, and the calculated areas swept out by each of the pie-shaped triangles proved to be exactly the same.

Kepler reasoned that there are invisible, interattractive tensional forces—i.e., zero-diameter "cables"—at work. He took a giant step toward describing the gravity existing between celestial bodies. Kepler's mind had discovered the nonsensorial and thus invisible relationships existing between celestial bodies, even though these interrelationships are not made evident by the behavior of any one of the bodies or parts of the system when considered only separately.

Brains can only discover via the senses. The interrestraint of those planets and the Sun could not be seen, smelt, felt, tasted, or heard. No one could see these zero-diameter tethers that exist between each of the planets and between each of those planets and the Sun. Mind alone, in contradistinction to the sensorially apprehending brain, had discovered invisible interrelations through the irrefutable data of scientifically observed and measured natural behaviors.

Kepler pondered, "If the diameter of the fibrous ropes or strings with which I accelerate the weights I swing around my head is progressively reduced by using ever thinner and fewer-fibered cords, eventually the cords will break." To think of a string of no thickness at all holding together bodies the size and weight of the Sun and the planets, and doing so across many millions and even billions of miles of space, is to consider physical interrelationships existing in Universe heretofore unapprehended by the senses, unanticipated by the senses and ergo imponderable by human brains.

Kepler had to think also about forces operating between and among groups of all the other planets and each planet, as well as between and among various groups of planets and the Sun. Because the planets orbit at different rates, from time to time they bunch together and at other times move far away from each other. When bunched, their combined local group mass produces greater pull on each of the individual planets than does their separated-from-one-another interpair pulls. Kepler realized that this causes the planets to move in elliptical orbits: ellipses being determined by a pair of restraining forces. Kepler had to think about, comprehend, and explain to the satisfaction of his own mind's functional integrity exactly how and why the solar system's intercoordinating tensions govern all these ever-changing planetary interrelationships.

Next, using the new cipher-implemented calculating possibilities, Galileo computed the rate of acceleration for free-falling bodies. He found them accelerating at a second-power rate of velocity in respect to the arithmetical distance traveled.

Isaac Newton soon became intensely and passionately driven to understand the invisible tension forces that Kepler had found operating across millions of miles of open interplanetary and intersteller space. Newton was very much advantaged by the experiences of others. He recognized, for instance, as must anybody living by the sea, that the full Moon brings with it much higher tides.

Newton sensed a vast body of water being pulled. A full Moon occurs only when the Moon, Earth, and Sun are in 180° alignment, with the Earth positioned in the middle. Newton saw how the combined 180° pull of the full Moon and Sun is very much greater than when the Sun and Moon are interangled at 90° to the Earth. Newton posited that the relative interpull between any two pairs of equiinterdistanced celestial bodies must be proportional to the respective pairs of products of their respective masses. Formulating his concept from Galileo's second-power-accleration discovery, Newton finally hypothesized that the rate of interattractiveness between any two celestial bodies varies inversely with the second power of the arithmetical distance intervening. That is to say, if you halve the distance between the two, you increase their interattractiveness fourfold. If you double the intervening distance, you quarter the interattraction. When asked, "What is gravity?" Newton would have had to reply, "It is nothing to which I can point. It is an interrelationship, existing only between parts."

From birth, it is given humans to desire to understand all the relationships of all their experiences, which is to say, to accomplish with the human mind that which brains cannot. Once in a great while human mind discovers one of those exquisite—only mathematically expressible—macro- or microcosmic interrelationships. Mind operates only and always synergetically.

Einstein's genius was synergetic. All genius is synergetic. All children are born geniuses, but most are swiftly degeniused by the power structure's educational system. In the guise of education, the system deliberately breaks up inherently holistic considerations into "elementary" topics.

Early in my 1927-initiated lifelong experiment, I realized that what we call a principle—for example, the commonly and constantly inter-varying rate of the mass interattraction of celestial bodies—could qual-ify as a generalized principle of science only if exceptions to the rule are never found. In other words, generalized principles are inherently eter-nal. Unfortunately, we tend not to recognize that which is eternal.

Eternity is invisible. The more persistently we think about it, the more we realize that when we say "no exceptions," we in fact mean eternal. Thus, we find human mind delving into, and sometimes dis-covering, eternally covarying interrelationships.

The human brain, on the other hand, always and only deals with the visible and temporal—i.e., special cases with beginnings and endings. Illogically, the brain seeks a cosmology with a beginning and an ending, whereas inherently eternal Universe has neither. The Universe could not have begun with a big bang.

All the big bang theorists—which is to say, the academic establish-ment—are illogical and brain-bound when it comes to questions of cosmology. Beginnings and endings are inherently special case.

The big question is where would all that energy for that primordial big bang come from, and wherefrom the space in which to stage that first big bang?

The speed of light was exactly measured at the opening of the twen-tieth century—186,000 miles per second, or approximately 5.87 trillion miles in a year. Astronomers adopted the light-year as the unit of dis-tance measure of astronomically observed bodies. Polaris, the North Pole star, is 470 light-years away from us observing it from Earth. In television parlance, it is a "live" show. Other stars are much farther distant, but they are all live (real time) shows, too, with their light taking from 4.3 years to many hundreds of centuries to reach us. Our Sun's light takes a mere 8 minutes to reach Earth.

Einstein operationally observed the Universe as a complex aggregate of nonsimultaneously occurring, variously directioned, variously inter-woven and overlapped, variously enduring events. I gave the name *scenario Universe* to Einstein's concept of Universe to distinguish it from a conventional single-frame picture, the concept of Universe fa-vored by Newton.

Nonsimultaneous scenario Universe is inherently without beginning and end. We shall delve further into Einstein's nonsimultaneous sce-nario Universe shortly. We introduce it here as the seventh cosmic, nonsensorially apprehensible interrelationship discovery.

Returning to our main line of thought, the other five of the twelve historically most outstanding of the human mind's cosmic-interrelationship discoveries are described in detail elsewhere in this book. To keep them in constant prominence throughout the reading of this book, however, I am listing them here:

The eighth discovery is Archimedes' principle of *similitude,* discussed in Chapter 1.

Ninth is *wisdom,* which I identify as the inherent acceleration in metaphysical evolution as a consequence of the cumulative, synergetic integration of only progressively acquired knowledge.

Tenth is mathematics, which of course includes Euler's topology.

Eleventh and twelfth are radiation and gravity, which always and only coexist. Disintegrative radiation and integrative gravity in symbiosis describe the elusive object of the quest for a "unified field." In a more poetic sense, these characteristics also identify love as being both shining radiation and all-embracing metaphysical gravity.

Love is the synergetic marriage of radiation and gravity.

Elucidating synergetics, we note that there is nothing in one atom per se that predicts that atoms will combine to form chemical compounds. One atom does not predict anything, let alone the existence of another atom or combinations of one known atom with an as-yet-unknown other atom.

Humans have witnessed quite naturally ("natural" because in an a priori synergetic Universe) that atoms combine. Beyond that, they have discovered the mathematical equations, but not the structural concepts of the manner in which atoms combine or thereby the existence of laws governing their intercombinings.

There is nothing in chemical compounds per se that predicts biological protoplasm. There is nothing in biological protoplasm per se that predicts camels and palm trees and the intercomplementary interexchange of the waste gases given off by them. There is nothing in the exchange of these gases that predicts galaxies and stars.

The greater complex is never predicted by the parts of the lesser complex. Therefore, I surmise that to learn anything you must start with the whole—with Universe.

Comprehension of the whole alone leads to discovery of the significant intercomplementary functions to be played by the parts.

To learn is to regain the cosmically comprehensive conceptual realization of our innate genius—to use our minds.

In view of this latter realization, I shall, in my further thinking, first and foremost address Universe.

First, I would like to examine all the generalized principles thus far discovered. They are not many. What, I ask myself, can I see regarding the whole inventory of those principles that I cannot observe by looking at only one principle at a time? Is synergy operative amongst the whole family of thus-far-discovered-by-science generalized principles: Ohm's law in electromagnetics, Avogadro's law and Gibbs's phase rule in chemistry, and Einstein's $E = mc^2$?

(Here I thought, Is Universe the synergy of synergies—i.e., $s^4 \times s^4 =$ synergy to the fourth power progressively fourth-powered? That speculative question, however, ventures beyond the scope of our present survey of verifiable scientific principles.)

Most impressive to me is the fact that, being eternal, none of all the thus far discovered generalized scientific principles has ever been found to contradict any other. All are interaccommodative. Many are interaugmentative.

When you and I use the word *design* in contradistinction to the word *random* we immediately include the concept of intellect, that sorting-out and recombining in intellectually preferred, synergetically interbehavioral pattern arrangements. Only intellect can formulate and express its design conceptionings mathematically—for instance, Einstein's mind-formulated and -expressed $E = mc^2$.

That the human mind has been designed to apprehend, to comprehend mathematically, and to express intellectually eternal-Universe design interrelationships and—even more—to employ these interrelationship principles in specially formulated objective-use cases as micro-macro-structures and mechanisms informs us that humans have indeed been designed and developed for cosmic-magnitude functioning. To discover whether this terrestrial installation of humans and their minds will lead to the fulfillment of this cosmic functioning, all human individuals are now entered upon their final examination.

Noting the disparate delays involved in light from celestial bodies reaching our cognition, Albert Einstein said that the observed Universe is an aggregate of nonsimultaneous, differently energized, differently enduring energy events, each with its own unique beginning and ending.

Einstein's worldview—that Universe is an aggregate of only over-lapping nonsimultaneous episodes—I have come to call "scenario Universe" because of its resemblance to an ever-changing film script with the threads of new comings and goings interwoven into a complex story.

Universe has no all-encompassing beginning and ending. In scenario Universe, beginnings and endings, births and deaths are local events. Big bang theorists, within the limits of their vision, ask only single-frame questions, such as this: I wonder what is outside the outside of Universe? The academic and scientific establishment, with credentials derived from Newton, conceives of Universe as a static structure, an object viewable as a whole, all at one time.

Nothing in a single-frame picture of a caterpillar tells you it is going to transform into a butterfly. There is nothing in a single-frame picture of a butterfly with spread wings to tell you it can fly or is flying. It takes many frames of a moving picture to tell you that it is flying; it takes millions of frames to give you any clue as to how it flies; and it takes thousands of scenarios to show why in the scheme of Universe the butterfly is designed to fly.

We wonder how it can be that nature develops a virus or the billions of beautiful bubbles in the wake of a ship. How does she formulate these lovely geometries so rapidly? She must have some fundamental, simple, and pure way of developing these extraordinary life cells and chemistries.

I discovered that the tetrahedron was at the root of the matter. I found that the tetrahedron was the minimum thinkable set which subdivided the Universe and that relatedness could be demonstrated. I found the organic chemist from an entirely different viewpoint discovering the controlling influence of the tetrahedron in vertex-to-vertex relation. I found the metallurgist half a century later discovering the fundamental role of the tetrahedron, but this time related edge to edge. Chemists and biologists, in their specialized disciplines, seem to be finding all the structuring of nature to be tetrahedrally configured.

I have found the tetrahedron to be the minimum structural system of Universe. The tetrahedron is basic to synergetic geometry. All polyhedra may be subdivided into component tetrahedra, but no tetrahedron may be subdivided into component polyhedra of less than the tetrahedron's four faces.

Fig. 2.1 is a drawing of a tetrahedron with its four vertices, four triangular faces, and six edges.

There are only three structural (omnitriangulated) systems in Universe. Of these three primitive structural systems, only in the tetrahedron are the vertexes free to plunge through their opposite triangle. In the other two innate structural systems, the octahedron and the icosahedron, the vertexes are prevented from plunging through to the opposite side of their structures by the existence of opposite structural components. But each vertex of the tetrahedron is exactly opposite a wide open triangular window.

At this stage in my exploration, I discovered that neither physics nor engineering had a description or definition of what they meant by the word *structure*. Structure in their fields of expertise has always been axiomatic—in other words, obvious for millennia. Obvious to physicists

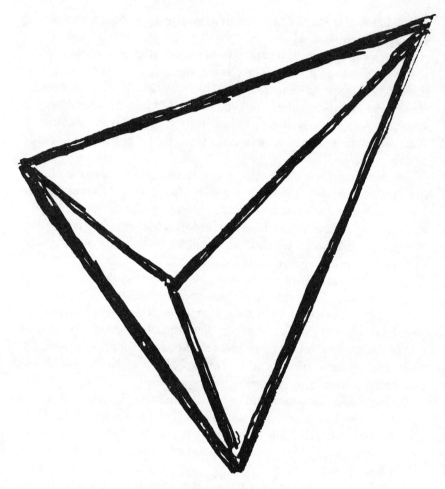

FIG. 2.1 *The tetrahedron.*

and engineers was, for example, the solidity of a block of marble or the rigidity of stone. I sought to discover how nature structures things.

In my search for a definition of structure I developed operational exercises that would eventually lead me to the experientially formulated generalization of tensegrity.[1] I constructed a necklace consisting of many 12-inch-long, ½-inch-diameter aluminum tubes strung on a Dacron cord (see Fig. 2.3). I found that the more tubes included, the more

[1] My contraction for "tensional integrity": The unified field model, constructed of struts and a discrete network of strings, integrating most economically both compressional and tensional elements into a whole system.

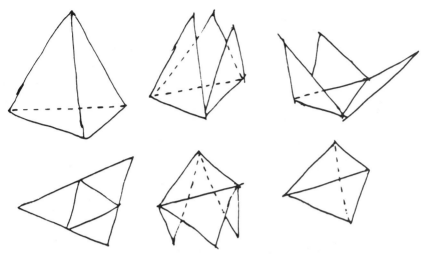

FIG. 2.2 *Unfolding a tetrahedron and turning it inside out.*

fluidly flexible was the overall necklace. Flexing this necklace neither altered its length nor bent any of the tubes. Clearly, flexibility was provided entirely by the tension joints. Because the tubes were not providing flexibility, I progressively eliminated them one by one, and the necklace became progressively more prominently angular. Finally, I had only three tubes remaining, and for the first time the necklace would no longer flex. It was a triangle with a triangular hole in it, the hole being larger than my neck. This experiment clearly demonstrated that the triangle is the only many-sided figure (polygon) that holds its shape, despite its three completely flexible corners. There was no two-tube necklace: it would not provide a hole for my neck to penetrate. The triangle was clearly the terminal case of polygon formation.

Since I found the pattern of my triangular necklace to be stable and since the triangular necklace that holds its shape consists of three separate push-pull, firmly shaped aluminum tubes and three flexible Dacron-cord corner-angle coherers, I formulated my definition of structure as ''a complex of events interacting to produce a stable pattern.'' Amplifying that interaction, I described ''a system whose component events are persistently interpositioned by a balance of forces of interrepulsion and interattraction.'' I found the necklace structure to be just such a complex of push-pull coherence integrity. I thus concluded that triangulation is essential to structure and that no necklace of more than three push-pull tubes is stable.

Since the minimum system in Universe, the tetrahedron, is entirely

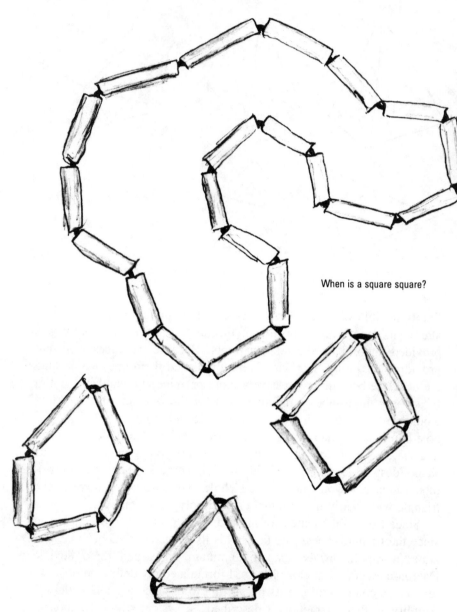

When is a square square?

FIG. 2.3 *Proving that the triangle is the only polygon to hold its shape and that thus its stability is fundamental to structure.*

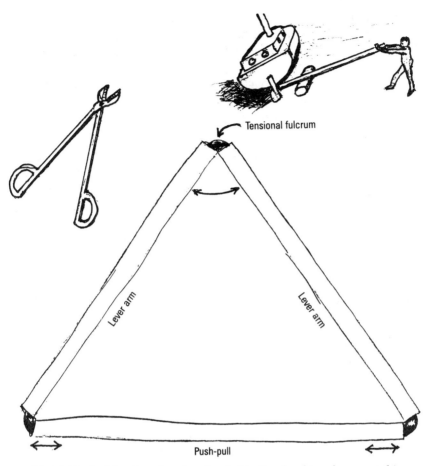

FIG. 2.4 *Each side of a triangle takes hold of ends of two levers, stabilizing the angle opposite with minimum effort.*

embraced by exactly four triangles and since the triangle alone produces a stable pattern, I concluded that *the tetrahedron is the minimum and simplest structural system in Universe.*

No wonder the tetrahedron and its contained octahedron (Fig. 2.7), together with its eternal, allspace-filling complementary octahedra, are the structural components of diamonds. No wonder Jacobus van't Hoff was the first chemist ever to receive the Nobel Prize, for his optical proof of the tetrahedral configuration of carbon. No wonder Paul Mac-Cready's *Gossamer Albatross* was light enough to be human-muscle-pedaled in its trans–English Channel flight; its structural components were formed of tetrahedrally stabilized carbon fibers, making it tensionally so strong for its weight that the 90-foot craft could be held up in one hand. The greatest wonder, however, is that tetrahedra and their significance are not included in college preparatory school curricula.

I NEXT UNDERTOOK TO DISCOVER why the three-aluminum-tubed, Dacron-cord-cohered triangle held its shape.

We discover that a pair of scissors consists of two edge-sharpened levers pin-fulcrumed one on another and that the longer the lever arms, the more powerfully they can cut. We discover that any two sides of our necklace triangle that are tensilely cohered are joined to one another at one end. We then discover that on the third side of the triangle we have a push-pull tube that is firmly seizing the outer ends of the two other tubes of the triangle, stabilizing the angle opposite it with minimum leverage effort. Thus does each push-pull side of the triangle stabilize its opposite angle with minimum effort (see Fig. 2.4). We find this minimum-effort characteristic to be consistent with all behaviors in nature, which always accomplish their patterning work with minimum effort.

The necklace triangle illustrates the principle of leverage advantage holding a complex of events motionless, in contradistinction to levers being used to move objects with minimum effort.

The reasons are many for the failure of physicists to include the tetrahedron and its component triangles in their exploratory strategy. Prime among these reasons is that physics has divorced itself completely from geometrically conceptual models, restricting expression of its explorations and findings exclusively to algebraically expressed formulae, with the assumption that calculations could always be translated into physical technology through the XYZ (axes) and c-g-s (centimeter-gram-second) coordinates of analytic geometry.

Being a science that is nonsystemic and committed to discovering only parts and guessing at the parameters that may be involved in their exploration, physics is intractably nonsynergetic.

Repeating ourselves for emphasis and confirming our experimental evidence with a different set of physical items, we note the following: As our two hands manipulate the ends of a pair of tied-together sticks (the sticks representing two vectors),[2] our hand motions flex the tied-together corner angle. A pair of sticks joined at one end articulate in the same

---

[2] In synergetic geometry, vectors exist only as energetic phenomena. A vector always represents a product of mass and the velocity of a given energy entity operating in a given angular direction in respect to a given axis of observation. Every energy event must have six vectors.

pattern as a pair of scissors. By the principle of leverage, the longer a pair of scissors' handles, the more powerfully they cut. The scissors' corner-in serves as the common fulcrum of the two levered-together handle extensions of the scissors' cutting knives. If a pair of scissor handles is open to an angle of 60° and if we then take a stick about the length of the scissor handles and fasten it to the handles' outer ends, the cross-tied stick (vector) will prevent the scissors from further flexing. This is accomplished with minimum effort because the ends of the cross-tied stick are tied to the outermost ends of the levers, thereby producing with the least effort the greatest leverage advantage in stabilizing the opposite shear angle. The stick holding the two lever ends apart thus produces a closed polygonal pattern—a three-flex-cornered triangle. The triangle is therefore demonstrated to be the minimum flex-cornered polygon (there being no two- or one-vector-edged polygons).

We have therefore demonstrated that each of any triangle's sides always stabilizes its opposite angle with minimum effort.

The triangle is the only flex-cornered polygon that holds its shape; ergo, it alone accounts for all structural shaping in Universe. Triangles do not, however, exist independently of systems. In synergetic geometry, the triangle is necessarily a very flat tetrahedron polyhedron, one with an almost negligible altitude (see Fig. 2.5C). The minimum system—the tetrahedron—has four flex-corners, four triangles ("windows"), and six vector-edge lines. Systems are independent in Universe and are therefore rotatably considerable. Systems always have two corners to serve as poles of system spin and other nonpolar corners in sets of two. For every set of two nonpolar corners, all structurally stable systems always have triangular windows in sets of four opposite the four corners and vector edges in sets of six—with no exceptions.

Of all polyhedra, only the tetrahedron can be turned inside out to become its own mirror image, or complementary opposite. To picture this, imagine any point of a flexible tetrahedron being pushed through its triangular base. The resulting figure is a mirror image of the initial tetrahedron, just the way a rubber glove turned inside out becomes its own mirror image. In this way, the tetrahedron demonstrates the inherent twoness of a system. Tetrahedra can be experimentally demonstrated to be *the* optimally economic, most comprehensive structurally integrated systems in Universe.

In time, the existence will be acknowledged of *both* the special-case physical, systemically considered Universe and the generalized metaphysical, comprehensive tetrahedron-Universe. Synergetics, the comprehensive geometry I have systematized, unlike all other systems of

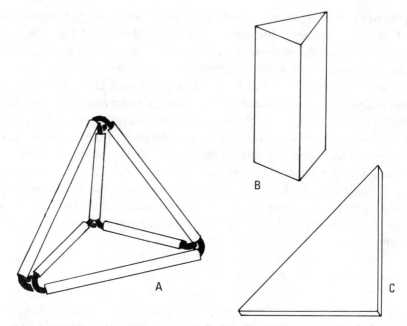

FIG. 2.5 *A seemingly independently existent triangle is always a four-cornered tetrahedron of minimagnitude altitude. A* is a four-flex-cornered tetrahedron; *B,* a prism; *C,* a flat piece of paper cut out as a triangle (in reality a prism of meager but geometrically significant altitude).

geometry, incorporates both the physical and metaphysical. (The metaphysical involves that which can be experienced but is independent of size and is weightless and energyless, i.e., qualititative rather than quantitative.)

Inherent twoness is all-pervasive in Universe. We recognize that concave and convex always and only co-occur. Because concave surfaces concentrate, while convex surfaces diffuse, reflectively impinging radiation, we find demonstrated at conceptual outset that concave and convex produce different energy effects, wherefore it is experimentally evident that unity is plural and at minimum two. There can be no oneness, for it would be undifferentiated from its background; it could be neither conceptualized nor described; it would have neither insideness nor outsideness.

There is another way to demonstrate the at-minimum-twoness of the Universe (*universe* means toward union, not toward isolatable oneness).

There is no such phenomenon as "oneness" possible in Universe.

One always presumes an other, in the same way that inside presumes outside and concave presumes convex.

The other at-minimum twoness of unity is the observer and the observed, and their union is the realization of life—in pure principle.

We can make a true model illustrating how the extra syntropic A Quanta Modules (which I shall describe shortly) produce the high-frequency interpulsing of the positive into the negative phase of Universe.

First, we make a triangle by welding together the ends of three 24-inch-long, 3/16-inch-diameter steel rods. We next take three high-tensile-strength, high-resiliency, interwoven-rubber-and-nylon-thread shock-cords and fasten one of each of their ends to the three corners of the steel-rod triangle. Then, taking out all loose slack, we fasten the three inner ends of the shock-cords together at the triangle's center of area.

Lifting the assembly and holding it before us with the triangular plane perpendicular to the floor, we now grasp the vertex formed by the knotted-together center of the three shock-cords (see Fig. 2.6).

We then thrust our hand forward and jerk it backward in swiftly alternating, successive movements. The inertia of the steel triangle keeps it in the same vertical position, while the shock-cords' flexibility permits us to push our swift forward-and-back motion of our fist in ever-deeper plunges and draws. This will be seen to be producing a succession of positive and negative tetrahedra. This means the tetrahedron is successively transforming its inside-out positive phase into its outside-in negative phase.

Geometrically, this is exactly what physicists find some atoms are doing as a constant characteristic of their existence. This phenomenon became the basis for the first atomic clock. Also this is precisely the

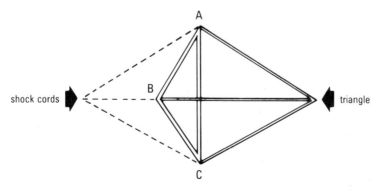

FIG. 2.6 *Pulsing of a tetrahedron as it turns itself inside out.*

way, in pure principle, time is introduced into an otherwise eternally timeless Universe.

Both recreational and academic mathematics have long been fascinated with what has been called four-dimensional geometry. Much speculation and puzzlement has centered on its amazing properties, such as exercises in magically crossing seemingly impenetrable surface boundaries or rotating such exotic forms as the hypercube. But difficulties have arisen in trying to model three-dimensional objects in the higher dimension by analogy.

Synergetics models such figures with ease. The tetrahedron is inherently four-dimensional, with four mutually related axes. Giving up our ages-old attachment to the right angle, we can now model four-dimensional figures and demonstrate their properties, thus showing that the fourth dimension is ordinary rather than exotic.

Where there is insideness and outsideness, there is a four-dimensional system. A flat paper triangle has insideness of the paper and outsideness. There is no surface apart from the object it bounds. There is no experimentally demonstrable one-, two-, or three-dimensionality. The tetrahedron, with its four planes of symmetry, is inherently four-dimensional. Four-dimensionality is the minimum: anything less is not a system and therefore cannot be conceptually considered.

There are only three primitive (i.e., pretime, presize, prefrequency of modular subdividing), most symmetrical structural systems in Universe:

A. *The tetrahedron,* with three equiangular triangles around each corner (four triangles total)
B. *The octahedron,* with four equiangular triangles around each corner (eight triangles total)
C. *The icosahedron,* with five equiangular triangles around each corner (twenty triangles total)

There cannot be demonstrated to exist a structural system with six equiangular triangles about each corner, because these six 60° angles add up to 360°, as do the angles around a point on a plane extending in all lateral directions to "infinity." Such a figure with six "equilateral" triangles around each point could only produce a plane forever unable to turn back upon itself to form a closed system dividing Universe into all Universe outside the system and all Universe inside the system, which is in fact the unique function of a system.

As a consequence, there are only three omnisymmetrical, triangularly structured systems in Universe: the tetrahedron, octahedron, and icosahedron. The Greeks revered these objects. Present-day engineers, academics, and physicists virtually ignore them. In developing my design science strategies, I sought to discover practical application of the design principles these systems embody and to design the way nature designs: with pristine logic and economy.

Life begins with awareness of otherness. All the other othernesses are always systems that have their own unique insidenesses and outsidenesses.

By this book's conclusion the reader shall have discovered the tools with which cosmologists, physicists, and mathematicians today confront the very biggest of questions in fields as abstruse as cosmology, quantum mechanics, and crystallography.

The reader will discover that the inexorable course of the gradual running down of the energy of the Universe—that is, entropy—is only part of the picture. Entropy has a complementary phase, which we designated *syntropy*. The reader will not only recognize these two phases of Universe but further will note and acknowledge that convex and concave modes are one way of picturing these phases. Convex may be viewed as the multiplication-by-division essential to quantum mechanics; in other words, from unity comes diversity. This convex phase represents the vectorially diffusive, entropic disintegration phase of Universe. Concave, on the other hand, is illustrated by simple multiplication and represents the syntropically integrative phase of only sumtotally regenerative Universe.

We recognize the tetrahedron, being simultaneously both convex and concave, to be thereby further qualified to serve as *the* comprehensive conserver of eternally regenerative Universe.

Reiterating, we note that tension and compression always and only coexist. Further, we have determined conclusively that gravity and radiation are this always-and-only-coexisting tension and compression functioning in their most inclusive macro- and microcosmic states. Central to the age-old search for a unified field theory has been the until-now-unsuccessful endeavor to reconcile gravity and radiation.

Of all that we classify as primitive (pretime and presize) imaginable closed systems, only the tetrahedron can be turned, or can turn itself, inside out. (See my definitive reference on the whole subject, *Synergetics,* Secs. 618.10 and 624.12.)

In *Synergetics 2* (1979), I continued my exploration of quantum mechanics' multiplication by division and by the inherent seven unique great circles of spinnability of all crystal systems and of all isotropic matrix embracements of symmetrical subsystems, and their successive multiplication by subdivision to produce not only the A and B modules but also what I call the S, T, and E modules. These great circles are also the same seven unique great circles of symmetry that are foldable into local-circuitry great-circle "bow ties," which are reassemblable into omnisymmetrical spheric systems, complexedly interweaving their spinning in great circles.[3]

Although I shall later provide a sensorial demonstration (see Fig. 3.3), to prove to myself that gravity is the inherent syntropic conserver of integrity in Universe, being twice as efficient as radiation and using only logic and the operational tools of synergetic geometry, I traced the following steps:[4] Multiplying only by division,[5] we proceed to bisect the six edges of the tetrahedron and most economically (that is, geodesically) interconnect those six bisection points (see Figs. 2.7 and 2.8). We then use those six symmetrically interrelated points as the three sets of poles of the three initial axes of rotation of the tetrahedron. All systems have cosmically inherent independent rotatability or spinnability. As we can clearly see in Fig. 2.8, these three rotations describe the octahedron as the first multiplication by division into one equi-vector-edged octahedron and four identical equi-vector-edged tetrahedra, with the central octahedron exactly equaling the sum of the volumes of the four corner-situate tetrahedra.

For as long as can be remembered academic science has embraced a cubical rather than a tetrahedron-based coordinate system.

One thing nice about the cube is that it neatly accounts allspace, without any other device. If we assess space as modern physicists do, with the cube as the measure of unit volume, we are using three times as much volume as necessary. If, on the other hand, we use the tetrahedron as the unit measure, we are practicing the economy that nature always follows in her designs.

When we use a cubical coordination system, we are being threefold inefficient. Because we are always dealing with physical experience and because physical experience in synergetics is nothing but structural systems whose edges consist of energy events whose actions, reactions, and resultants consist of one basic energy vector, the cube therefore

---

[3] See *Synergetics*, Sec. 954.10; and *Synergetics 2*, Secs. 986.440, 986.550, where I describe my discovery, naming, and listing of these synergetics modules and their compounding.

[4] Technical though this demonstration may seem, it requires little formal background in mathematics.

[5] In synergetics, as in quantum mechanics, we have multiplication only by division, because we begin with the whole (unity), and this unitary "Universe" expands only through progressively differentiating out—that is to say, subdividing.

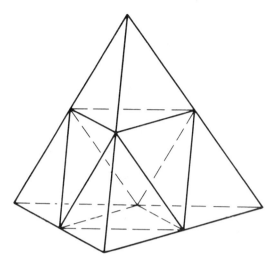

FIG. 2.7 *The three great-circle-spun square planes exactly bisecting the tetrahedron in three symmetrical ways.* The three triangular-system-formed subdivision-aspect squares are *ACBD, AEBF, DECF.* Note the primary tetrahedron and the secondary internal octahedron, and only then are the implied square cross sections of the octahedron apparent as tertiary derivations of the primary structural system, the tetrahedron. There is no single-plane, omni-equal-angle, equal-edge "square" structural integrity in Universe. Squares and cubes are always and only tertiary derivations of prime vectorial structuring systems.

requires three times the energy to structure it than the tetrahedron does. We thus understand why nature must use the tetrahedron as the unit of energy, as its energy quantum—because it is three times as efficient. All the experiments in physics show that nature always employs the most energy-economical tactics.

When we attempt to use tetrahedra as the "building blocks" of a coordinate system, we quickly discover that they will not fill allspace. The octahedra and tetrahedra must pack *together* to fill allspace, with no intervening pockets of space.

Tetrahedra and octahedra agglomerate to fill allspace; they complement one another. To the individual looking for a monological explanation, this synergetic model would be unsatisfying. To the physicist, who recognizes complementarity as a basic principle, this method of accounting would be rational and very satisfying indeed. The complementary, allspace-filling grid of tetrahedra and octahedra is given the name *isotropic vector matrix* in synergetics because of the unique property of the grid being composed of equal length elements and being

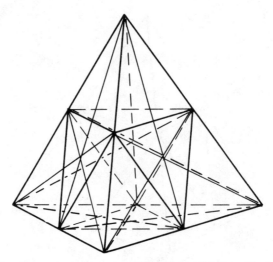

FIG. 2.8 *The six great-circle-spun subdivisions of the tetrahedron—what I call the A and B Quanta Modules.* All regular polyhedra (other than the icosahedron and the pentagonal dodecahedron) are composed of fractional elements of the tetrahedron and octahedron. These elements are known in synergetics as the A and B Quanta Modules. They each have a volume of $1/24$ of a tetrahedron (see *Synergetics,* Secs. 910–916). This illustration shows the six great-circle-spun subdivisions of the regular, primitive tetrahedron into its twenty-four A Quanta Modules and of the contained octahedron into its forty-eight A and forty-eight B Quanta Modules by the further symmetrically spun four great circles of unique spinnability of the four axes of the eight opposite regular triangles of the tetrahedron-contained octahedron (see *Synergetics 2,* Sec. 987).

everywhere the same. The grid may be thought of as a schematics of the contact points when spheres are closest-packed.

In synergetic geometry, this allspace provides a rational, numerical, and geometric framework upon which to model nature's own most economical coordinate system. This framework I identify as the isotropic vector matrix, which fills allspace with a grid composed of tetrahedra and octahedra.

The elegantly simply structure upon which I base this system is composed of tetrahedra and the octahedra formed inside each tetrahedron by connecting the midpoints of the six edges of the tetrahedron.

Rational, numerical, and geometrical values derive from (a) parallel and (b) perpendicular halving. The thirding and physical isolation of the prime number three and its multiples is only an inadvertent consequence of the three-way, symmetry-imposed, perpendicular bisecting of each of the tetrahedron's four triangular faces. The parallel method of tetrahedral bisecting has three axes of spin and ergo three equators of halving; and the perpendicular method of tetrahedral bisecting has six axes of spin and ergo six equators of halving. Halving and its inadvertent thirding introduces the twenty-four A Quanta Modules.[6]

This discussion leads us to the A and B Quanta Modules, which, I intend to show, become the rational, numerical, and geometrical units of all geometries and of all crystallography.

To reiterate this most important discovery, the tetrahedron is spinningly fractionable in several ways:

1. The successive spinning of each of three great circles fractionates the tetrahedron into an internal octahedron of volume 4 surrounded by four small tetrahedra each of volume 1. How do we know that? Because, when the edge module of a system is 2, its triangularly modulated surface is $N^2 - 2^2 = 4$ and the system's tetrahedral volume is $N^3 - 2^3 = 8$; therefore, a tetrahedron with edge module 2 has a volume of eight regular tetrahedra. Subtract the four corner tetrahedra from the overall tetrahedron volume of 8 and the octahedron that remains is $8 - 4 = 4$ volumes; i.e., the octahedron has a volume of four tetrahedra of the same vector-length edge modules (see Fig. 2.7).

2. The six great circles fractionate the tetrahedron into twenty-four A modules. The six great circles are the extensions of the tetrahedron's six edges over and downward beyond the vertexes as the perpendicular bisectors of the two successively encountered equiangular triangles (see Fig. 2.8). The six great circles are spun on two sets of three axes each, running between the three half-altitude points of the two adjacent pairs of triangular faces of the tetrahedron (see Fig. 2.7).

3. Finally, four great circles are spun about the four axes provided by the perpendiculars from the tetrahedron's four apexes, impinging perpendicularly upon the center of area of their four opposite triangular faces.

The three and the four and the six great circles taken all together fractionate the original omnienergy quantum tetrahedron of physical Universe into ninety-six A modules and forty-eight B modules—i.e., two A modules for every B module in Universe. These modules are the two basic units from which, I contend, all rational, numerical, and geometrical values derive, as well as all phenomena of crystallography.

---

[6] A Quanta Module: A fundamental structuring element of synergetic geometry, one-sixth of a quarter-tetrahedron, which will be more fully described further on.

Synergetics provides an alphabet of working units with which diverse fields of study can be reconciled without resorting to awkward, irrational, or fractional values.

Because the A modules are foldable into their tetrahedral form from only one whole triangle, energies entering them inherently bounce reflectively around within them. For this reason, A modules conserve their energy receipts (see *Synergetics*, Sec. 913).

Because the B modules' tetrahedra are each folded together from four different triangles, the energies entering the B modules are reflectively dispersed from them (see *Synergetics*, Sec. 916).

4. Algebraically described, we have:

$$(+) \cdot (+) = (+)$$
$$(-) \cdot (-) = (+)$$
$$(-) \cdot (+) = (-)$$

The A Quanta Module occurs in nonnestable pairs: the syntropically conserved, self-regenerative energy of the A+ module $(+)$ and the syntropically conserved, self-regenerative A− module $[(-) \cdot (+) = (-)]$. The two A's have a constant in Universe $(-)$, whereas the alternative left and right winging of the inherently entropic B modules operate singly, left-handedness producing a negative proclivity, and right-handedness, a positive proclivity. Therefore,

$LB = (-) \cdot (-) = (+)$ and $RB = (-) \cdot (+) = (-)$

Therefore,

$LB = (+)$ and $RB = (-)$

Therefore,

constant $(A+) \cdot (A-) = (-)$

Therefore,

constant A pair $(-) \cdot (LB+) = (-) =$ gravity coherence

Therefore,

constant A pair $(-) \cdot (RB-) = (+) =$ radiation

Therefore, we have twice as much gravity (i.e., coherence) as we have radiation.

Gravity or coherence = syntropy
Radiation or disintegration = entropy

Therefore, the Universe is twice as powerfully integrating as disintegrating (i.e., twice as powerfully syntropic as entropic).

Extrapolating from this demonstration, we can surmise that one-half of the integrative forces of physical Universe rule over the disintegrative forces. The other (excess) half of the integrative forces are invested in a constant oscillation between the positive and negative modes of the tetrahedron.

For the first time humans have been able to have a conceptual picture of a local electromagnetic wave disturbance. Unlike other attempts at linear or planar models of this phenomenon, synergetics provides a multidimensional wave-propagation model (the "jitterbug") and its description of the rotation of the tetrahedron between its two phases within a cubical framework.

This phenomenon generates all electromagnetic wave motions, effecting both a positive and negative phase of Universe. The negative phases, being disconnects of eternity, produce both time and eternal evolutionary transformation.

Time intervals, thus, are split-second black-hole glimpsings of the negative phases of Universe. The second set of A Quanta Modules permits time to stretch out diverse, overlapping episodes into the non-simultaneity of eternally regenerative Universe.

It is this time-lapsing capability of the syntropic A modules that permits the momentarily "negative" lapse that we human, time-embraced phenomena think and speak of as life. Without time, there is no what-we-think-of-as-life.

Life begins as a special-case episode of our awareness progressively discovering the always-present otherness of "plural-unity"[7] and its multiplication by further dividing into a complex of overlappingly episoded experiences always terminating daily with sleep, from which we emerge each time to start a new set of awareness-of-otherness dreams. No one has ever been able to prove that the human who awakened was the same human (who may always be dreaming) who went to sleep yesternight.

Experienceable unity is plural and at minimum two. The system's inherent insideness and outsideness, its concavity of insideness and convexity of outsideness, coexist in pure principle, where we cannot have one without the other. Since concave concentrates impinging radiation and convex diffuses the same radiation, concave and convex do not perform the same function; ergo, the minimum otherness experience of life's awareness is a system unto itself whose insideness and outside-

---

[7] Unity is plural and at minimum two.

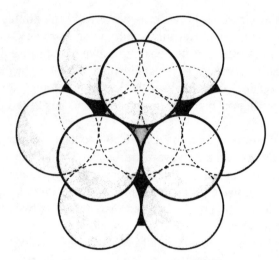

FIG. 2.9 *Spheres closest packed twelve around one.*

ness demonstrate that unity is always plural and at minimum two. Zero corners plus zero faces equals zero edges plus two.[8] Universe is two.

IN SEARCH OF THE PRINCIPLES UPON which nature structures Universe, I further identify what I call the *coupler* as the uniquely asymmetric (or only polarly symmetric) octahedron, which is comprised of the many in-the-same-space-reorientable combinations of the quarks. I shall return to the coupler after I describe how I arrived at its discovery.

Nature's coordinate system, I determined, fundamentally consists of a matrix of tetrahedra and octahedra, which together fill allspace. To model this isotropic vector matrix, I first observed how spheres stack. I pictured identical cannon balls stacked in the way by nature they tend to stack most economically.

Spheres always and only closest pack tangentially with twelve spheres around one (see Fig. 2.9).

When spheres of unit radius are closest-packed, there are two kinds of spaces intervening: the concave octahedron and the concave vector equilibrium spaces. These two can be assembled edge to edge with one another to produce a ''continuum'' of allspace-embracing, closest-packed, unit-radius spheres. Such an assembly will not have whole spheres on the outer surface of the assemblage, but instead will have

---

[8] Using Euler's formula, which will be discussed further on: number of corners plus number of faces equals number of edges plus 2.

FIG. 2.10 *Rhombic dodecahedron.*

only concave surfaces with the appearance of a mass of hardened clay covered by the concave impressions of half-shells of long-dead clams.

There exists a polyhedron with twelve diamond faces. It is called a rhombic dodecahedron (see Fig. 2.10). Structurally stable rhombic dodecahedra closest pack with one another and, in doing so, actually fill allspace, as do the nonstructurally stable cubes only theoretically. (*Theoretical* means "assuming you are God and are playing the game of inventing the rules of the game of the experience called life.")

The centers of volume of the closest-packed rhombic dodecahedra are congruent with the centers of volume of the closest-packed unit-radius spheres whose radii are the same as the twelve radii of the rhombic dodecahedra, which radii are the perpendiculars to the twelve mid-diamond faces' centers of the allspace-filling rhombic dodecahedra.

We thus discover that the twelve diamond-faced, allspace-filling rhombic dodecahedra are the allspace-filling "domains" of each of the closest-packed unit-radius spheres whose closest packing is also that of all atoms.

A sphere fits neatly inside the rhombic dodecahedron, with each of the dodecahedron's twelve mid-diamond faces tangent to the enclosed sphere at the same twelve points of tangency of the twelve spheres closest packing around one another. Since the rhombic dodecahedra fill allspace while containing the spheres when closest packing together in the same twelve-around-one pattern of the spheres that are each tangent to the mid-diamond faces' centers, we can understand why rhombic dodecahedra are the domains of spheres.[9]

[9] In synergetics' omnitopology, spheres represent the omnidirectional domains of points. Each of the lines and vertexes of polyhedrally defined conceptual systems has a unique areal domain and volumetric domain. To give just a few examples, the volumetric domain of an external face is the volume defined by that external face and the

When the vector-edged regular tetrahedron's volume is 1 (i.e., unity), the vector-edged octahedron's volume is exactly 4, the symmetrically interjoined, positive-negative tetrahedron's eight-cornered overall cube aspect has a volume of exactly 3, and the rhombic dodecahedron has a volume of exactly 6; i.e., the volumetric unit of allspace filling is exactly 6. Allspace unity equals 6. Unity is plural and volumetrically at minimum 6.

Polyhedron	Volume
Tetrahedron	1
Octahedron	4
Cube	3
Rhombic dodecahedron	6

Synergetics' constant unit of length is the edge of the tetrahedron and, therefore, of the isotropic vector matrix, which, we recall, identifies the allspace-filling, omnidirectional grid composed of alternating tetrahedra and octahedra, neither of which fills allspace without its complement (more precisely, its dual).

To make a cube with a volume of exactly 3 hold its shape, a tetrahedron must be inserted into it. The tetrahedron's edges form the diagonals of the square faces of the cube. In conventional academic science's *XYZ*, 90°, square- and cube-coordinated system with its N^2 squaring and N^3 cubing, the cube's edge N is unity. In synergetics the tetrahedron's edge N is unity. When we use synergetics' vector constant as the edge of the cube instead of as the diagonal of its faces, the volume is 3.5339 versus the volume 3 of synergetics' vector diagonal cube.

The vector-edged cube's volume is the irrational number 3.5339 +. This 3.5339 + cube is the vector-edged cube that physics illogically, encumberingly, and slavishly uses and has always used as the unit volume in the centimeter-gram-second and *XYZ*-coordinate system of academia's energetic mensuration. Using its volume as the standard unit volume for the entire hierarchy of primitive symmetric polyhedra makes them all awkward, irrational values. The measuring system used by business and industry and taught in every university science department is thus a mishmash of awkward, cumbersome values. Aesthetically inclined students are repelled by the irregularity and disorder.

When allspace-filling rhombic dodecahedra are closest packed and

---

center of volume of the system; and the surface domain of a polyhedron's external lines is inherently four-sided and is the area defined by the lines most economically interconnecting the centers of area of each of the polyhedron's faces with the ends of the lines dividing those faces from one another.

the long diagonal of their diamond faces is vector-lengthed, there are exactly twelve around any one. Any two closest-packed rhombic dodecahedra have their respective common diamond faces congruent with one another.

If we interconnect the centers of volume of any two adjacent rhombic dodecahedra with the four corners of their common diamond interfaces, we produce the only polarly symmetric octahedron (see Fig. 2.11). It is this figure that I named the *coupler,* for it couples not only the centers of volume and centers of energy of the closest-packed, allspace-filling rhombic dodecahedra but also the centers of all the closest-packed unit-radius spheres and thereby of all closest-packed atoms. Couplers inter-couple centers of energy—the nuclei—of all closest-packed unit-radius (i.e., equiwavelengthed) atoms.

Consisting of twenty-four modules, the coupler's volume is identical to the volume of the tetrahedron. It is here that we identify what nuclear physicists have named *quarks,* theoretical subatomic particles that carry fractional charges and such fanciful characteristics as "upness," "downness," "strangeness," and "charm."

The coupler always consists of eight *mites,* or quarks—the three right-angled isosceles tetrahedra consisting of two energy-conserving A modules, one of which is inside out of the other, and one energy-dispensing B module of either the inside-outness or outside-inness phase.

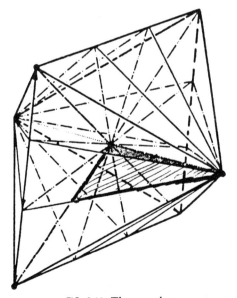

FIG. 2.11 *The coupler.*

There are all together two internal alternative interarrangeabilities of the mites' two A modules and one B module—$A+$, $A-$, $B+$ left wing; or $A+$, $A-$, $B-$, right wing—all within the same overall, allspace-filling, right-isosceles tetrahedron.

These two interchangeable energy-conserving and -dispersing behaviors correspond exactly to those that nuclear physicists attribute to quarks. The mite's geometrical space domain has two noteworthy internal module arrangements producing two uniquely different energy-conserving effects and one energy-dispersing effect, as does, also and exactly, the quark's.

The number of different interrearrangements of the mites within the coupler's 8 tetrahedral receptacles is

$$\frac{N^2 - N}{2} = \frac{8^2 - 8}{2} = \frac{64 - 8}{2} = \frac{56}{2} = 28$$

This 28 we multiply by the twoness of internal mite rearrangeability of the mite's 2 A and 1 B modules, giving us 56 rearrangements of the same total energies inter-energy-proclivities of each coupler. Each spherical atom has twelve couplers linking its center to the centers of its surrounding neighbors:

$$\frac{N^2 - N}{2} = \frac{(12)^2 - 12}{2} = \frac{144 - 12}{2} = \frac{132}{2} = 66$$

Ergo, we have $66 \times 56 = 3{,}696$ different energy-holding and -dispersing proclivity variants of interassociatability of each atom always within identical, superficially static, atomically closest-packed array space.

With 3,696 potentially unique interrelationships within the first shell, and $(42^2 - 42) / 2 = 861$ multiplied by 3,696 for the second shell, and $(92^2 - 92) / 2 = 4{,}186$ multiplied by 3,696 for the third unique shell of any nucleus, and $(162^2 - 162) / 2 = 13{,}041$ multiplied by 3,696 for the outermost shell of closest-packed limit uniqueness of any given unit-radius, symmetrical, closest-packed, nucleated system, then $3{,}696 + 861 + 4{,}186 + 13{,}041 = 21{,}784$.
Further,

$$\frac{21{,}784^2 - 21{,}784}{2} = \frac{474{,}520{,}872}{2} = 237{,}260{,}436$$

This number is the number of unique subnuclear componentions of each nucleus with which you have to play.

Synergetics provides real-world understanding of interarrangeabilities of subatomic particles, which is to say, a more sophisticated understanding of subatomics than that of the nuclear physicist whose favorite tool is the atom smasher.

ALTHOUGH I HAVE GONE INTO THIS subject in a certain amount of detail, what I have intended to demonstrate is simply that the framework of synergetic geometry makes possible the discovery of many varieties of subatomics all within the same seemingly static space.

Through the use of synergetic geometry, then, particle physics, which is one of the more abstruse and esoteric areas of frontier theorizing in science, falls within the grasp of the ordinary individual, allowing him or her to consider, to model, and to puzzle over it. Synergetics uses simple models based on a few basic modules that fit together in the most logical possible ways. Synergetics uses whole numbers, completely eliminating all irrational, imaginary, and irresolvable numbers and complex formulae. It is amazing that technology has been able to produce what it has, considering the obstacle presented by current scientific conventions in the field of geometry and measurement. The scientific and academic establishment still cowers in the Dark Ages imposed by human power structures many centuries ago. The dawn of scientific civilization is yet at hand.

In recent correspondence with a nuclear physicist, I urged him to continue his intensive study of synergetics as presented in my two volumes on synergetics. I gave him, however, a strong warning that I could not guarantee that other physicists would accept his inferential deductions and identification of them with the findings of the conventional *XYZ,* c-g-s calculus of academia's subatomic explorers.

I told him it would be a multibillion-dollar savings to society each time he successfully identified one of the millions of now "colorful" and "strange" subatomic particles through use of synergetics' A, B, S, T, E modules and the myriads of their extendabilities.

Government-financed, private-enterprise-exploited atomic accelerators and their kindred producers spend about a billion dollars per subatomic particle discovered, whereas I have firmly established and classified all that they have or ever will soon discover, and vastly more, only at the cost of living expenses for self and family during my fifty-four-year program.

These are my own half-century-ago discoveries, comprehensively published together for the first time in *Synergetics.* As discoverer, original graphic illustrator, and namer, I ask all explorers in the field of synergetic geometry for respectful use of my system of naming when setting out to identify the significant interrelationships of the vast variety of subnuclear rearrangement arrays.

I HAVE OFTEN STATED THAT BY *Universe* I mean the aggregate of all humanity's consciously apprehended and communicated (to self or others) experiences.

All the individual experiences in this aggregate of omniexperiences cannot be simultaneously recalled. They can be recalled only in systemic increments. The individual, systemic recallability from memory of many experiences—some rapidly, some slowly—suggests possible omnirecallability in extended time of the entire memory-banked collection of the majority of individuals' unique experiences.

Among the total accumulation of special-case experiences of all humanity we sometimes discover interrelationships existing which display a mathematical orderliness which always and forever demonstrates absolute consistency in its mathematical interrelationship. These exquisite interrelationships we identify as only mathematically expressible generalized principles. An example of such a generalized principle is the discovery that the number of unique interrelationships of any given number of entities is always $(N^2 - N) / 2$ (see *Synergetics*, Sec. 227). Another example is the law of similitude, which showed shipbuilders that doubling the length of their freighter allowed them to carry eight times as much cargo.

The aggregate of generalized principles derived from the aggregate of all humanity's consciously apprehended and communicated special-case experiences can be said to express most exactly and economically what we mean by Universe.

Eternally regenerative scenario Universe is an aggregate of principles.

To qualify as principles, they must be exceptionless. When stated positively, *exceptionless* means eternal.

The synergetic complex of omniinteraccommodative eternal principles is inherently weightless and changeless, ergo metaphysical.

Metaphysical Universe and its component principles are omniweightless and only metaphysically expressible. The metaphysical principles are one and all so absolute that their interoperative behaviors become mathematically tune-in-able and tune-out-able.

A very small range of this vast spectrum of tune-in-ableness becomes sensorial to humans. The sensorially apprehensible principles are what we call the *physical* aspects of omnimetaphysical Universe.

What distinguishes the physical from the metaphysical is not what both the casual and trained observer might note: solidity, opacity, hardness, or heaviness.

The physical is either the tune-in-able or interferable, coincidental, interceptible, special-case, sympathetic resonance of substances and/or electromagnetic frequency ranges of the human senses within the comprehensive metaphysical frequency and wavelength spectrum.

Solids are themselves only wave complexes. They are the superficially deceptive microaggregates which defy differentiating resolution into their myriad separate parts by the unaided eye.

EVERY ONCE IN A WHILE IN THE 1950s and 1960s, philosophy scholars and others in the academic world would say to me, "If you are ever confronted by Professor Weiss at Yale, some of your basic theories are liable to be dismantled." On my appointment in 1968 to a Hoyt fellowship at Yale, Professor Weiss and I were encouraged by the students to appear together on the Yale University television station. Though we had not as yet met, we accepted the invitation readily and independently.

Professor Weiss was a widely known, distinguished professor. I met him for the first time in the television broadcasting studio. The station program director seated us opposite one another at a stout wooden table. On the director's signal that the recording of the program was commencing, Weiss thumped the table resoundingly with his fist, saying, "Don't tell me that this table is not solid."

I replied, "How can you see me over here, defiantly glaring through what are obviously solid spectacles?" To which the professor opened his lips to reply—his mouth fell open—but no words came.

I proceeded to explain that glass is an aggregate of very high frequency atomic events and that a good analogy would be an alignment between Weiss and me of a number of rows of airplane propellers rotating so fast that none of their blades can be seen. If he reached his fist toward me, an invisible solid, like his eyeglasses, would thump his fist or cut it off.

Because the speed of the propellers is directly coupled to their controlled-speed motors, it is possible using gears to time and aim a battery of machine guns to shoot bullets between the spinning blades, as I described in Chapter 1 of this book. I explained to him that the speed of light is so swift that it can readily pass through the circular-motion patterns of his glasses' whirring electrons. By making the glass lenses a little thicker, the distance the photons of light must travel (at 186,000 miles per second) is increased enough to permit mild interference with the gyrations of the electrons of the atomic components of the glass. As a baseball is angularly redirected just a bit by a batter's foul-tip, this refractive direction-changing of the light-photon passage makes possible lens correction of our physical vision equipment.

Professor Weiss asked the studio to cancel the program and walked out. Thus, we discover how seemingly "hard realities" may be only mathematical differentiations of frequency and angle, operative in pure principle.

Synergetics, alone among generalized system theories, models Universe in its many-splendored effulgence so completely and pristinely using only frequency and angle.

Since there are no things, no solids—only events operating in pure principle—and since no events touch other events in Universe, Universe is coordinatingly cohered, formed, and transformed only tensionally, repulsively, electromagnetically, and gravitationally; even the event "electron" is as remote from its nucleus as the Earth is from the Moon, in terms of these regenerative systems' respective diameters.

The term *solid* has come in recent years to mean subvisible behaviors, as in the development of solid-state physics. Science evolved the name *solid-state physics* when, immediately after World War II, the partial conductors and partial resistors—later termed transistors—were discovered. The phenomena were called "solid-state" because without human devising of the electronic circuitry, certain traces of metallic substances accidentally disclosed electromagnetic pattern-holding, shunting, route-switching, and frequency-valving regularities, assumedly produced by the invisible-to-humans atomic complexes constituting those substances.

Further experiment disclosed unique electromagnetic circuitry characteristics of various substances without any conceptual model of the "subvisible apparatus." Ergo, the whole development of the use of these invisible behaviors was conducted as an intelligently resourceful trial-and-error strategy in exploiting invisible and uncharted-by-humans natural behavior within the commonsensically solid substances. The addition of the word *state* to the word *solid* implied regularities in an otherwise assumedly random conglomerate.

What I have discovered goes incisively and conceptually deeper than the blindfolded (Dark Ages) assumptions and strategies of solid-state physics—whose transistors' solid-state regularities seemingly defied discrete conceptuality, scientific generalization, and kinetic schematizing. Synergetics provides the subatomic explorer with a roadmap leading to discrete conceptuality, scientific generalization, and a schematic for further exploitation.[10]

Synergetics discloses the seven sets of great circles (four on the vector equilibrium and three on the icosahedron) that produce all the fractionating and polyhedral facetings of all crystallography.

Synergetics also shows that all four of the great circles of the vector equilibrium (VE) transverse the twelve vertexes of the VE and that those vertexes are the same points of interconnection of unit-radius spheres in closest-packing. I later point out that energy charges always follow the convex surfaces of spheres and that ergo those four sets of great circles constitute the Universe's only "railroad tracks" for energy transmission via atomic agglomeration. I also show that the twelve vertexes of the icosahedron could be pump phased ("jitterbugged") into congruence with the VE's twelve vertexes.[11]

Synergetics also discloses the foldability of each of these seven great circles into local bow-tie-like patterns, which act as local-circuit shunts and are reassemblable into whole-sphere integrities. Totally assembled, they reconstitute the whole great-circles patterning of the completed spheres. They demonstrate thereby that these great circles may act as local information-shunting and -holding circuits. I also show that the icosahedron and its three sets of great circles may serve as a compre-

---

[10] Adjuvant's Note: In the mid-1980s, new discoveries were made about the carbon atom, despite the fact that carbon has been subjected to more study than all other elements put together. Because these discoveries, by Dr. Harry Kroto of the University of Sussex and Drs. Robert Curl and Richard Smalley at Rice University, Texas, demonstrate principles first outlined in synergetics and in the work of Buckminster Fuller, this entirely new family of carbon molecule clusters were named the fullerenes, of which Buckminsterfullerene, with the most stable molecular structure in the family (C_{60}), is a special case. Although carbon exists in forms as diverse as diamonds and graphite, which vary only in their atomic arrangement, the exceptional stability of this new hollow-cage structure, according to Dr. Kroto, has shed a totally new and revealing light on several important aspects of carbon's chemical and physical properties that were quite unsuspected and others that were not previously well understood. Despite the fact that scientists had assumed they were familiar with all forms of carbon, this whole new chemical family with symmetric molecules made up of hexagonal and pentagonal arrays of from 28 to 540 or more carbon atoms was an unexpected revelation. The fullerenes, according to recent speculation, could be the source of a whole new class of chemical compounds. A remarkable property of C_{60} is that it appears to form spontaneously, which fact has particularly important implications for particle formation in combustion as soot and as stardust in space. These molecules floating in interstellar space could play a critical role in planet formation and be responsible for mysterious spectral lines emanating from stars, an observation that has puzzled astrophysicists for decades (see *Science*, Nov. 25, 1988).

[11] "Jitterbug" is my pumping model, made of wooden struts and rubber connectors, that shows the circumferential and radially covarying states that a polyhedron undergoes—for example, as it contracts from a VE to an icosahedron. See Fig. 6.77.

hensive information shunter-holder of even greater capacity. The icosahedron's system of thirty-one great circles is capable of releasing and routing its information most economically in uniquely preferred contact directions, hinting at the possibility of molecular-level computer technology. My studies show that it is possible to understand the discrete energy shunting and holding patterns at the molecular level.

Synergetics, further, discusses such new-era computers latent in the atomic world now to be mathematically reached and employed at the most exquisitely microcosmic minitude.

Synergetics discusses the secondary sets of great circles of both the vector equilibrium and the icosahedron and the part they can play in computer systems.

Getting back to my counsel to subatomic explorers and my nuclear physicist correspondent in particular, I recommended that they study all the tables of calculations of spherical and planar triangular subdivisioning of all the secondary great circlings of Universe—with the dimensions being given of all the central angles (arcs and chords) and surface angles both polyhedronally and spherically, in the Appendix of Tables starting at page 477 of *Synergetics 2*. These tables comprise one key to my strategy of eventually arriving at an entire cosmic system that accommodates all possible transformations through only-whole-number accounting.

In addition to the subjects already discussed, I submitted to him another important extension of my material on comprehensive strategies for mathematically generalizing and both omnirationally and only-whole-number accounting of the entire cosmic system to accommodate any and all of the nonsimultaneous intertransformings and interexchangings of finite but nonunitarily conceptual Universe. I had no qualms about the importance of pursuing this strategy because I innately knew that nature only worked with whole-number, rational accounting in her myriad designs.

In this connection—that of a comprehensive all-embracing whole-number accounting system—I asked him please to read and study *Synergetics*.

Just before *Synergetics* went to press in 1975, I discovered what I call the Scheherazade Number, of which this seventy-one-integer number is the latest version:

$$2^{12} \cdot 3^8 \cdot 5^6 \cdot 7^6 \cdot 11^6 \cdot 13^6 \cdot 17^4 \cdot 19^3 \cdot 23^3 \cdot 29^3 \cdot 31^3 \cdot 37^3 \cdot 41^3 \cdot$$
$$43^3 \cdot 47^3 = 616,494,535,0,868,49,2,48,0,51,88,27,49,49,00,6996,$$
$$185,494,27,898,13,35,17,0,25,22,73,66,0,864,000,000$$

This supreme seventy-one-integer Scheherazade Number can also be presented in columnar form in order to disclose a surprising number of symmetries. This number embraces a minimum n^3 number of all the prime numbers involved in evolving all trigonometric functions and all the surface and volumetric spherical system intertransformings of synergetics.

Using this number as the number of divisions of circular unity, with the comprehensivity and speed of computers, it is possible to rework the calculations of all the trigonometric functions. If, as I predict, all the results are in whole-rational-number increments (without any decimal fractions), we can then assume that all scientific calculations could be reworked with this comprehensive dividend base.

As noted before, quantum mechanics is founded on the assumption of the total of energy in Universe being unincreasable, wherefore all multiplications of its investments in physical work can only be accomplished by division of the finite whole—what I call "multiplication by division." If our seventy-one-integer Scheherazade Number is employed as the comprehensive dividend, all calculations should always be resolvable in whole rational integers.

The last set of references introduces you to what I am confident are the cosmically primitive properties of number that govern all physical behaviors. Thus, we have an octave system consisting of four positive and four negative numbers and one empty, twixt octaves zero: 1 adds 1, 2 adds 2, 3 adds 3, 4 adds 4, 5 subtracts 4, 6 subtracts 3, 7 subtracts 2, 8 subtracts 1, 9 neither adds nor subtracts (its effect is zero).

The last set of references also introduces you to the fact that the product of multiplying the fourth, fifth, and sixth prime numbers—7, 11, 13, which superstition has stigmatized as the "bad luck" numbers—produces the 1,001 of the historic *Thousand and One Nights*. As these last references also show, these particular numbers continue to produce left and right half-mirror symmetry and, when compounded with the first three primes, produce very impressively rememberable patterns of numbers.

If you use the seventy-one-integer Scheherazade Number as the number of subdivisions of a great circle, you can recalculate the sines and cosines only for each degree of a circle of 360°. Having done so, if you find all the resulting 1° increments to be whole (fractionless) numbers, needing no "rounding off," then we may assume that our seventy-one-integer divided may quite possibly accommodate holistically and rationally our scientific calculation at the extreme reaches (both micro and macro) of humanity's instrumental search.

Nature employs only whole atoms. Nature employs only whole systems.

Of course, the Scheherazade integer increments will be too big for ordinary use, but they may well be reduced in size by first lopping off the same number of zero tails from each and all of the results and thereafter reducing them all by successive common divisors. All of the foregoing can be computer-remembered and may lead to a whole new world of scientific discovery of *absolute* interproportioning.

We may well find a much lower comprehensive dividend than the Scheherazade Number to be adequate to all cosmic-energy behavior accounting in whole rational increments. But, in any case, only rational numbers need be used—in other words, numbers that can be expressed as ratios of whole numbers. For example, nowhere in *Synergetics* is it necessary to introduce irrational numbers such as pi, which is approximately 3.14159265 + and irresolvable. Rather than futilely carrying pi out to ever more million decimal places and wondering when nature decides to "round off" her calculations, I assert and maintain my strategy of only calculating with rational, whole numbers—confident that my strategy is the one by which nature abides.

In all my thinking which I have been sharing with you, it has been my working premise that:

1. Life begins with independent individual awareness of otherness.
2. Independent individual awareness must have its own unique *outsideness* and *insideness* which makes it an individual system.
3. Awareness occurs always and only within the physical brain.
4. Image-I-nation is always and only stimulated from outside the brain by information supplied through the nervous system by the feeling of internally or externally located pain, touching, tasting, smelling, hearing, seeing, or possibly by other infra- or ultrasensorially tunable electromagnetic frequency receptors.

All the evidence of all science's experiential findings, whether read from invisible-magnitude evidence instruments or from comprehensive visual observations of complexes of facts, must ultimately be apprehended always and only within the human brain's image-I-nation, the omniscience-coordinating and systematically omniframeless, TV-like systemic conceptualizing.

We may therefore say with scientific certitude that all of science's experienceable evidence is always and only an imaginative experience. All experiences are imaginable only as conceptual systems and are always geometrically, topologically, and vectorially expressible as generalized or special-case system experiences.

All generalizations are metaphysical and eternal—i.e., independent of time. All special-case experiences begin and end and are therefore temporal. Brains always and only deal with special-case temporal phenomena. Minds alone deal with the only mathematically expressible eternal interrelationships of Universe.

Mathematicians speak of numerical generalizations as "empty sets"; thus, an empty set of five is the generalized prime number "fiveness," whereas five people or five fingers are special cases. Even more specialized cases are you and me and our very special individual cases of five personal right- and five left-hand fingers.

Employing only four imagination-experienceable (i.e., physically evidencible) cosmic-event loci and only six structural, push-pull vectors to omniintegratingly interposition those four event loci, thereby omniempoweringly and embracingly employing all the available energy of Universe in the fewest and simplest ways, the primitive tetrahedron accomplishes the conceptual defining of the simplest omniclosed system configuration of Universe, which system quantumwise inherently divides all the Universe into:

1. All of the Universe outside the special-case, tuned-in, four-corner-event loci defining the considered special-case tetrahedral convexity system

2. All of the Universe inside the special-case, tuned-in four-corner-event loci of the considered[12] special-case tetrahedron's concavity

3. All of the untuned-in generalized Universe outside, ultrafrequenced and ultrairrelevant to the special-case, tuned-in considered Universe

4. All of the untuned-in, generalized Universe infrafrequenced and infrarelevant to the considered special-case tuned-in Universe

5. All the remainder of the for-the-moment, special-case, tuned-in, and exclusively considered Universe, which does the dividing of the macrocosm from the microcosm

6. All the remainder of the generalized Universe dividing the generalized macrocosm from the generalized microcosm

7, 8, 9, 10, 11, 12. The six negative Universe phases of the tetrahedron's inherent transformability from its outside-outness to its inside-outness

---

[12] Considered = con-sidered, from *sidus* = star. *Considered* means the interjoined array of stars of which we are thinking—the constellar array that taken together is the tetrahedron or other "considerable" concept.

All twelve of the above quantum-multiplying-only-by-dividing are further quantum divisible by the purely metaphysical principles of topological aspect abundance-inventories of vectors, and time-occasioned angle and frequency interference actions, reactions, and resultants.

Because the most economical tetrahedron accomplishes definition of the simplest, omniconsidered, omni-Universe differentiating and integrating structural principles capable of demonstrating closed withinness and withoutness system-integrity, we thereby conceive of the individual episodes of the only-overlappingly-episoded scenario Universe as being minimally structured, ergo, with the tetrahedron, because it is the simplest conceptually primitive structural system able to define a closed-episode withinness and withoutness system integrity.

The tetrahedron is the simplest minimally componented metaphysical generalization of systematically thinkable conceptualization.

A synergetic system inherently and conceptually divides all Universe into all of the Universe outside the system, all of the Universe inside the system, and the small portion of the Universe that constitutes the system that does the dividing. That part of the Universe outside the system is itself divided into all that is inherently relevant to the dividing system and all that is irrelevant to the tuned-in, considered system. Also, that part of the Universe inside the system is divided into that which is relevant to the system considered and all that which is not presently tuned-in as relevant to the system considered. As we approach a system, we come from the macroirrelevant into the macrorelevant and then into the system itself. Next we penetrate into the microrelevant zone and then into the microirrelevant.

The system considered could be the discretely tunable nucleus and its family of microcosmic, allspace-filling particles, which themselves are frequency modulatable and ergo subject to discrete system tune-in-ableness. The nuclear system is the turnabout phase of the Universe, at which the inbound considerations terminate and the outbound considerations ensue.

The most recently exposited quantum mechanics is predicated on the most updated concept of nonsimultaneous, complexedly overlapped, only special-case beginnings and endings of individualized occurrences within the unique episodes of scenario Universe. In such a Universe, beginnings and endings are only local-in-time inceptions of syntropic gatherings overlapped with terminal entropic dwindlings of systemic entities—for instance, of the progressive, always mathematically orderly gatherings of electrons in atoms, and of the mathematically orderly gathering of atoms in molecules, and the gathering of molecules in protoplasmic cells, and the gathering of cells in biological fibers, and the gathering of separately begun and ended fibers into threads, and the gathering of only-overlapped-separately-begun-and-ended threads in the strands, and the gathering of strands in ropes, and of the various ropes interspersed in the complex events of human environments, and so on.

Moreover, it can be scientifically demonstrated that all physical systems are continually giving off energies—a process we call entropy. Owing to each of the local Universe system's unique periodicities, these energies are randomly expended in respect to other systems. Thus, various localities of the physical Universe are expanding and expending energies in an increasingly disorderly manner. But fundamental complementarity requires that there must be other localities and phases of Universe wherein the Universe is reconvening, collecting, and concentrating in an increasingly orderly manner as a complementary regenerative conservation phase of Universe, thus manifesting a turnaround from increasing local disorder to locally increasing order, from entropy to syntropy.

The surface of the planet Earth seems to be just such a place.

Scenario Universe involves only a constant sum total of nonsimultaneous energy events and an only overlappingly aggregated complex of syntropic systems, which in the case of the photosynthesizing biology of planet Earth are predominantly recovering the entropically lost energy of predominantly entropic systems, such as those of all the stars. With the inherent syntropy of the planet Earth's biological photosynthesizing of orderly molecules out of the random, entropically broadcast energy receipts, and the consequent photosynthetically evolved biological hydrocarbons, which combine in an orderly manner with other orderly organic atoms of the stardust and other celestial entity receipts, altogether integrating in a planetary aggregation, the Earth's ecology further syntropically organizes into the omniintercomplementary, ecobiological complex of orderly designed biological species and special-case, individual biological organisms, all together depositing energy into fossil fuels against a multibillions-of-years-from-now, star-igniting functioning. The stored-up fossil fuels on Earth, in other words, will someday enable this planet to become a star. And likewise, as humans reach the Moon and beyond, lifeless spheres may someday become more Earthlike.

Uniquely syntropic amongst the biological species are the minds of humans, which have the semidivine capability of discovering and objectively employing the only-mathematically-expressible, thus-far-in-history-discovered aggregate of generalized principles.

The significance of terrestrial ecology's antientropic functioning was coincidentally discovered and independently published individually by myself and Norbert Wiener as constituting the most comprehensive and incisive antientropy manifest in Universe. In 1951 I rechristened the negatively expressed "antientropy" *syntropy*. This observation should have, but did not, terminate the assumption of astrophysicists that there exists no reversing of the star-manifest entropy (or "heat death") of Universe. Scientists and philosophers alike have continued to ponder and search afar for the possible existence of "entropy violations"— missing here, close at hand, our obviously regenerative Earth, a virtual terra incognita.

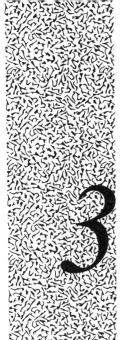

EINSTEIN

I HAVE SPECULATED A VERY GREAT DEAL about the significance of Einstein and his epistemology. I have written and lectured about him for many years.

Only in the context of my direct experiences with Einstein do I have a right to talk of him. Because he told me directly that he approved of the way I analyzed his teleological processing of experience into thought and the latter into systemic formulations and formulae, I have had great confidence in continuing to do so for the past forty-eight years.

In 1930 Einstein wrote an article for the *New York Times Magazine,* "The Cosmic Religious Sense: The Non-Anthropomorphic Concept of God." In this article, Einstein wrote that Kepler, Galileo, and other scientists who had been labeled heretics and cast out by the Roman Catholic church seemed to him to be much more imbued with a faith in the exquisite intellectual orderliness and sublime integrity of Universe than were the topmost Roman Catholic clergy. Einstein said, "What faith in the orderliness of Universe must have inspired Kepler to spend all the nights of his life alone in contemplation of the stars." Einstein reasoned that humans cannot undertake that kind of total isolation unless they are deeply inspired and have absolute faith in, and a clear sense of, the integrated significance of that orderliness. This integrity Einstein spoke of as God. It was a nonanthropomorphic god—not shaped like humans or any creature whatsoever. Einstein described the demonstration by humans of such faith in the orderliness of Universe as constituting the cosmic religious sense.

Deeply inspired by that article, I started writing my first major book in 1933. I named the book *Nine Chains to the Moon* because I had found that a head-to-foot chain of all human beings on planet Earth would reach back and forth between the Earth and the Moon nine times. I hoped the "nine chains to the Moon" concept might encourage locally preoccupied humans to dare to think more globally and even more cosmically.

Nine Chains to the Moon began with what I called a "tentative cosmic inventory" of the 1933 limits of what science knew—which was very, very little—about both the macro- and micro-Universe and its intermediary operational behaviors. I carefully checked far and wide with scientists regarding inclusions in the cosmic inventory. I faithfully listed everything considered important regarding all the experientially obtained information on the macrocosmic-microcosmic physical-phenomena limits thus far attained. When *Nine Chains to the Moon* was to be reissued forty years later, I looked at that inventory again and was shocked at the paltry limits of 1933 technological attainment and the meagerness of pre–World War II scientific knowledge.

Einstein's essay "The Cosmic Religious Sense" was published as Chapter 2 of my book, with the permission of its author and publisher.

For my third chapter, I considered how a man like Einstein, with that kind of philosophy, thinking as he did, happened to develop the concept of relativity and how he came to his many other preeminent conclusions, such as his revolutionary equation $E = mc^2$.

Looking into the facts of Einstein's everyday life, we find, for instance, that he was not only a schoolteacher but also for quite a while an examiner in the Swiss patent office.

Having taken out a great many patents of my own, I am aware of the process of writing a patent claim. One starts with a general review of the most advanced state of that particular invention's art and then discloses what one has discovered as a technical means for solving a problem, which technical means has never before been conceived of, realized, or proven.

As a patent examiner in Switzerland, a country that had developed the world's best timekeeping devices and led the world in the production of clocks, watches, and chronometers, Einstein must have read a vast number of patent claims on timekeeping devices. Implicit in these invention claims was the fact that nobody had ever found an absolutely accurate timekeeper. Inventors might develop improved accuracy, but none could attain perfection—which is true to this day. All this must have led Einstein to realize that Newton had to be entirely wrong in assuming a perfect uniform time to be a phenomenon instantly, simultaneously, differentiallessly operative and absolutely accurately observed throughout all the Universe. I felt sure Newton's error in assuming a universally and simultaneously uniform time impelled Einstein to start thinking along different lines.

I then concluded that a man who had Einstein's kind of philosophy and Einstein's kind of patent-examiner-of-timekeepers experiences would naturally think a great deal about relativity of nonsimultaneously and always differently viewed time experiences.

Next, I posited to myself, as best I could, not knowing Einstein personally at that time, how and why he might have come to formulate his various working assumptions; I had my own intuited explanation of how he formulated his epoch-initiating concepts.

Chapter 4 expressed my realization that in the world of science, when somebody does make a great breakthrough, "the Academy" is slow in officially acknowledging that breakthrough and acquiesces only when convinced by experiential evidence. Only then does the scientist's discovery or concept appear in school textbooks.

After the discovery is incorporated into textbooks, the new concept has finally arrived and enters the thinking environment of the everyday educational system. At this point, technological innovators commence thinking and speaking in terms of the new knowledge and its possible significance in solving old problems. Following the invention of an appropriate new artifact, there is a time lag before an industry adopts that invention. I call this lag the gestation rate, after its analog in biology. Only after a gestation period do the various new technological tools and goods springing from an invention change the everyday socioeconomic climate and physical environment. Eventually the altered environment induces everybody to think spontaneously like the scientist whose reasoning led to the original breakthrough.

My 1927 studies in techno-invention lags relative to various fields of scientific exploration and industrial endeavor showed me that it would take at least fifty years for Einstein's thinking to become everybody's "frame-of-thought" reference. My working assumption was that in time Einstein's theory would prove experimentally to be correct. It was not until 1942 that Einstein's principal formulation was proven to be both valid and accurate, when the Enrico Fermi pile showed that $E = mc^2$ correctly predicted the energy to be released from a given mass.

I had assumed that what Einstein was thinking would in due course be proven and would begin thereafter to affect everyday life. The current concerns about nuclear warfare and disarmament and the questioning of nuclear power plant safety somewhat confirm those predictions. For everyone to think the way Einstein did, however, we must rid ourselves of the Dark Ages concepts still taught in schools. Newtonian physics must be put into historical context, rather than propounded as the final word.

Based on that prognosticating logic in 1934, in my manuscript for *Nine Chains to the Moon*, I tried to conceive what our planetary life would be like if society indeed began to live in ways comprehensively consistent with Einsteinian thought.

My publisher had agreed to publish my book only because a great, successful author had recommended it. Six months after submitting the manuscript, I received a letter from the publisher's editor in Philadelphia that said, "You have three chapters on Einstein, and we've looked up the list of all the people that understand Einstein and you're not on it. In fact, we can't find you on any list. As a consequence, we think we must not publish your *Nine Chains to the Moon*. We want to avoid being a party to scientifically untenable speculation."

Dismayed, but being young and a bit fresh, I wrote back to the publisher, "Dr. Einstein has just now come from Europe to Princeton, New Jersey. Why don't you send my typescript to him and let him be the judge?" I had no hope that they would do such a thing, but nine months later, my Woodmere, Long Island, home telephone rang. The call was from a Dr. Fishbein, who said, "I live in New York City. My friend, Dr. Albert Einstein, is coming in from Princeton this weekend to stay with me. He has the typescript of your book and would like to talk with you. Do you think you could come in?" Obviously, I accepted.

On Sunday evening, I entered Dr. Fishbein's large Riverside Drive apartment. A number of Einstein's friends were already there. They were seated around the walls of an enormous drawing room. Einstein was seated at the far end of the room. As I was presented to him, I felt mystically moved. My reverence for him was such that I seemed to sense a halo above his head. He immediately arose, excused himself, and led me to Dr. Fishbein's study. On the desk of the study, under the lamp, I saw my typescript. We sat down on opposite sides of the desk. Einstein said he had read my typescript and found no fault. Better than that, he said that he liked the way in which I explained how he happened to come to think as he did and how he had formulated that thinking into his relativity theories and equation. Then Einstein said, "I'm advising your publisher of my approval of your explanation of my formulations."

Next he spoke to me about the fourth chapter of my book, which I called "$E = mc^2$ = Mrs. Murphy's Horse Power." I will never forget the gentle way in which he said, "Young man, you amaze me. I cannot conceive of anything I have ever done having the slightest practical application." He then went on to explain that he had made all his formulations in hope that they would be useful to cosmologists, astrophysicists, and physicists. He had no idea that any of his concepts and formulae would have any everyday practical applications whatsoever.

That meeting with Einstein occurred in 1935. Four years later Otto Hahn and Fritz Strassmann in Germany discovered theoretical fission. They conveyed their secret to their scientist friends in America. We all know what happened subsequently. Einstein was assumed by scientists to be the only one amongst them with sufficient credibility to convince Franklin Roosevelt that the Germans were working on the atom bomb and that the United States had better take advantage of this information and do something—fast.

Roosevelt responded with support and funds for the Fermi pile experiment, which proved Einstein's equation to be correct. Fermi's pile led to the Manhattan Project, the Alamogordo secret deployment, and the subsequent atom bombing of Hiroshima and Nagasaki.

Having heard Einstein say what he did, I could imagine how he felt when he learned what the first "practical application" of his thinking had wrought in Japan. His intimates saw how deeply it depressed him to the end of his life.

I am convinced that Einstein was very importantly stimulated by the work of Albert Michelson, who was intimately involved in accurate speed-of-light measurements. Nothing could have impressed Einstein more than the fact that Michelson had accurately measured that speed in a mile-long vacuum tube—and done so for all the different types of radiation.

As tiny children we assume spontaneously that our five senses are exactly time coordinated. Then comes the surprise one day when we see somebody pounding on a fence post at a distance from us and we realize that we hear the pounding after we see it happening. We thus realize that at least two of our senses are not reporting simultaneously.

Isaac Newton, along with all but one of the seventeenth-, eighteenth-, and nineteenth-century scientists, assumed it to be in physical evidence that light permeated Universe instantly and that therefore time also instantly permeated Universe. To them, Universe was both instantaneous and simultaneous.

Olaus Roemer, royal astronomer and mathematician to the King of Denmark, took exception to this thinking. In 1675 he observed eclipses taking place on the satellites of the planet Jupiter. The displayed lags between appearances of eclipsing shadows on the satellites and on the planet itself, their respective interdistancing and revolution rates, and their respective orbitings, convinced Roemer that light, like sound, has a unique speed and is not a "no-time-at-all," instantly everywhere phenomenon.

Not until the speed of light was scientifically measured on board our planet Earth 230 years later did other scientists pay serious attention to the phenomenon—but not to Roemer.

Roemer had excellent astronomical data about the distances intervening at any given time between the Earth, the Sun, and Jupiter, and between Jupiter and its satellites, as we have already noted, making it possible for him to calculate the speed of light, which he proceeded to do. His results closely approximated the measurements achieved by Dr. Michelson and his associates during a series of tests conducted throughout the first third of the twentieth century. Furthermore, Michelson's measurements and remeasurements with increasing exactitude were applied to the entire visible light spectrum and the invisible electromagnetic wave ranges, showing that all radiation, visible or invisible, has the same speed when unfettered in a vacuum.

Einstein could not have been more intuitively excited by this measuring. A number of other scientifically proven phenomena also stimulated Einstein's synergetic consideration: the Brownian movement is one; blackbody radiation and the discovery of finite photons of light are others. Since all radiation as energy unfettered in vacuo has the same speed, Einstein hypothesized that all slower-speed phenomena must be the result of the 186,000-mile-per-second radiations given off by the myriads of radiant-energy concentrations interfering with one another and tying themselves into knots to produce microcosmic inter-event-pattern systems that we humans identify superficially as matter. Einstein related everything to the speed of radiation, giving rise to his basic assumption that this speed is the norm of cosmic energy unfettered in vacuo. Einstein's norm will eventually replace Newton's norm of inertia, which he states in his first law of motion to the effect that "a body persists in a state of rest or in a line of motion except as affected by other bodies." Newton's norm of "at rest" is the accepted baseline norm of all twentieth-century economic and technological performance charts. The baseline's "at rest" means "no change." All our present techno-economic charts register changes occurring in time and at rates of change in respect to Newton's baseline of no change at all.

Einstein's norm of 186,000 miles per second assumed that when any less-than-norm speeds are manifest, energy is interfering with itself locally to tie itself into "knots," which are local holding patterns that humans speak of as matter. Einstein portrayed energy as existing in these two states: a slow phase of local self-interfering patterns, called matter; and a normal phase, as a spherical wave front traveling at the speed of light. This became the epistemological basis for Einstein's $E = mc^2$, where c^2 is the speed of light to the second power—which is mathematically derived from the fact that the area of the omnioutward, spherical surface wave growth of all radiation must be the second power of its outward linear-radius velocity.

To Newton, the norm of life and of the physical Universe in general was rest. To him, it seemed abnormal to have anything in motion; thus, death was the normal state. Newton reasoned that it took energy to put something in motion, as with a human muscle rolling a stone, and that the energy quickly became dissipated by friction, returning the stone to its norm of rest. Like all the classical scientists of his time, Newton subscribed to the concept that all energetic systems continually dissipate their energy, disposing of it in ever more disorderly ways. In later years this concept became known as the second law of thermodynamics and was given the name *entropy*.

Newton's norm of at rest, or no change, still governs the art of all graphic charting of evolutionary events—technical, economic, or social—when plotted against calendar or clock time. Newton's no-change norm forms the baseline of all such charts. The progressive magnitudes of change in evolution or development are posted vertically above the Newtonian baseline for the successive rightward calendar- or clock-time measurement.

Since the magnitudes of most historic, technologic, economic, or social performances are progressively increasing, our charts of development show an ever more abnormal trending of human affairs, suggesting an acceleration into verticality—which is utter abnormality—or "race schizophrenia."

If, however, Einstein's norm of 186,000 miles per second is substituted for the "motionless" norm of Newton's baseline, we have only to revolve 90° clockwise the charts plotted on the Newtonian norm. We will see then that humanity in its earliest and greatest ignorance was tailspinning into extinction, but, in the aviator's terms, is now progressively "pulling out into straight-and-level flight" (see Figs. 3.1 and 3.2) at the newly realized-to-be-normal speed of electromagnetic radiation's information transmission—i.e., 186,000 mps.

Newtonian reality was locked into the pre-Kepler, pre-Galilean Dark Ages. As already noted, Newton's gravitational conceptioning showed that the interattractiveness of any two given celestial bodies, as compared to any other pair of bodies a given equal distance apart, is proportional to the product of the respective pairs' masses and that the magnitude of their interattractiveness varies inversely as the second power of the arithmetical distances intervening. This conception of Newton's was developed (1) from Kepler's extraordinary realization and proof of a zero-diameter tensional restraint (line of force) operating between celestial bodies of unlimited magnitude and at apparently unlimited distances apart (for instance, the planet Pluto, a solar-captured comet-planet that orbits the Sun once every 247 years, is over four billion miles from the Sun), and (2) from Galileo's measurement of the rate of acceleration of free-falling bodies toward Earth, which was the second power of the arithmetical distance traveled. We must correct our cosmic-phenomena comprehension to accommodate the realization that since there is no *up* or *down* in Universe, there are no falling bodies. Instead, there are only nontouching, individual celestial bodies, large and small, whose normal motions of continual interpositioning are manifesting the Newtonian law.

It is important here to realize that both Kepler and Galileo started their reasoning with the observed fact that the Universe is always and everywhere transforming; these motionful transformings, as with all generalized scientific principles, are inherent in eternally regenerative

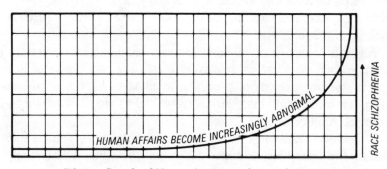

FIG. 3.1 *Graph of Newton's norm of "no change."*

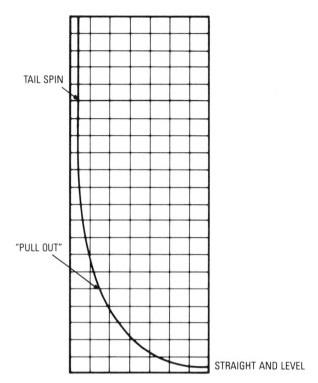

FIG. 3.2 *Graph of Einstein's norm of 186,000 mps.*

Universe. It was observing Brownian movement on the microcosmic level that triggered Einstein's working assumption that constant motion was the norm for physical Universe.

When the interference pattern of two or more motion events occurs, it does not mean that either or both events come to rest—i.e., that motion ceases. A chip may fall on somebody's shoulder, but this does not mean that the individual on whose shoulder the chip landed is motionlessly at rest or that the planet on which the individual on whose shoulder the chip landed is motionlessly at rest or that the planet on which the individual dwells is at rest or that the electrons of which all matter is comprised stop orbiting around their far-within, atomic, nuclear-event complexes.

Though unaware of atomic nuclei, Galileo and Kepler apprehended motion as an eternally operative principle, not as something that has to be initiated by something else. Universe is a nonsimultaneously differentiating complex of interference-occasioned relative rates of energy events, reflections, refractions, fractionations, formations, deformations, transformations, intertransformations, interaccelerations, interdecelerations, expansions and contractions, associatings and disassociatings, cotravelings and diametric separatings (radiation).

Newton used only his brain, which deals in special-case, time-dimensioned phenomena that ask for beginnings and endings for everything, including Universe. Kepler and Galileo used their minds and found relationships existing eternally between cosmic phenomena. What Galileo discovered was the rate of local interaccelerating of the eternally inherent cosmic acceleration. Multiplying a number by itself is second-powering. Galileo identified this second-powering as accelerating acceleration.

Galileo's falling body was in effect a very small celestial body being interattractively accelerated by a relatively enormous celestial body, the planet Earth. Newton's statement of his first law of motion (''a body remains in a state of rest or in a line of motion except as affected by other bodies'') makes it clear that he had not escaped from the nonenergetic, conceptual paralysis of plane and solid geometries, the pseudoscientific conceptual geometry tools of the Dark Ages. The laws of motion of Kepler and Galileo, however, were based on experientially proven measurements of cosmic behaviors, demonstrated a half century before Newton's hypothetical formalization and publication of their results.

Much of Newton's work must be considered political tour de force— British ''science'' in support of Great Britain's claim to the leadership of world science as backed by the world's supreme power structure. The fact that Newton's *Principia* develops all the geometry of his celestial mechanics by trigonometry, and not by calculus, casts a shadow of doubt upon his claim that he developed the calculus before Gottfried Leibniz as an invention of necessity to make possible his scientific discoveries.

Newton's failure to recognize and acknowledge Roemer's speed-of-light discovery postponed world science's academic thinking being advantaged by that knowledge. (It was knowledge of light and its speed that surely catalyzed Einstein's epochal thinking over two hundred years later.) Although Roemer had superbly, scientifically discovered that light has a speed, Newton ignored completely this finding when he published his *Optics,* thereby lessening the historic significance of that work. Newton seems deliberately or perhaps subconsciously to have sought to disregard Galileo's discovery of the second-power rate of variance of the celestial bodies' mutual interattraction in respect to the intervening arithmetical distances, rationalizing to himself that Galileo was dealing only with locally falling bodies and not with generalized interrelationships among celestial bodies. By limiting the Galileo discovery to a very-special-case local phenomenon of falling, which could only occur within the imaginary conditions of a static, infinitely extended lateral-plane, center-of-Universe, up-and-down, heaven-and-hell world, it was seemingly left to Newton to make the *big* scientific generalization.

We note in examining documents of the period that before Newton, Galileo had identified, numbered, and named his own laws of motion. Newton cast these aside as he nominated his first, second, and third laws of motion.

Newton's being knighted for his work in the management of the British mint suggests that his scientific work and the great reputation it brought him may have been affected by the interests of the behind-the-scenes power structure at that time promoting a full sovereign-scale British world empire, to be realized a century later with the Battle of Trafalgar.

I am dwelling on this Newtonian epistemology in order to emphasize the fact that Newton's norm of at rest left it to Einstein finally to emancipate the scholarly world—and thereby, in due course, world society in general—from its overwhelmment by the ignorant impotence of the Dark Ages, which had been established and maintained for seventeen hundred years by the might, cunning, and ruthless treachery of an absolutely selfish, deliberately self-misinformed world power structure in the form of the imperialism of the Roman emperor-popes.

My thinking has been inspired and accelerated by Einstein's insights. His written work has refined my speculative epistemology. This occurs to such an extent that I often find myself explaining Einstein beyond any record of his thoughts concerning the matters discussed, yet feeling spontaneously confident that the way I am conceiving on his behalf is so in accord with what I have learned of his way of thinking as to justify my extrapolations of his thought. In such a way do I also often unconsciously give Einstein complete credit for my own direct, experientially exciting epistemological excursions—using what I am confident were the thought exercises he used.

From time to time my "thinking out loud" in public addresses absent of prepared notes or outline becomes in fact real-time thought exercises integrated with experientially informed conceptioning.

Within Einstein's sphere of thought, I am most anxious to identify his assumption of a 186,000-mile-per-second inherent cosmic velocity norm with his concept of a nonsimultaneous and only partially overlappingly episodic scenario Universe. Such a finding would verify, support, and clarify Kepler's and Galileo's omniinteraccelerating, inherently and eternally intertransforming, nowhere-and-nowhen-ever-intertouching, exclusively intertensioned Universe.

To better understand this omniinteraccelerating cosmic concept, we must recall the following:

A. Universe is inherently resonant. Resonance is a complex of inter-transformative frequencies of miniintertensioned systems.
B. The inherent resonance of Universe is caused by nature's never pausing at, and only forever transiting, exact equilibrium.
C. The union of Universe is a differentially complementary regen-erative-production wedding of inherently, uniquely prime numbers 1, 2, 3, 5, 7, 11, 13, and all their successive primes. The prime numbers are numbers divisible only by themselves and by 1, representing in synergetics unique system behaviors.
D. The prime numbers impose an eternal disquietude—transformative adjustings and omniintertensioned resonances eternally interaccelerating.

Professor Robert Goddard, of twentieth-century rocketry fame, realized that Newton's gravitational interattractiveness variance law explained how a bicycle lying on the surface of the Earth is speeding around the Sun in tandem with the Earth at 60,000 mph, wherefore the bicycle's additional acceleration by a pedaling rider makes it accelerate faster than the Earth and, together with the mass of its pedaling rider, causes it to start to leave the Earth (as is demonstrated by chasm-jumping motorcyclists) and ergo to become dynamically stabilized, with the bicycle and its rider's integrative center of mass cotraveling just outward from, and a little bit faster than, the Earth's surface.

Goddard saw that with sufficient additional acceleration an Earth-cotraveling object would part company with the Earth and, if sufficiently accelerated, could reach its own orbiting distance outwardly from the Earth, at which distance and speed the attractiveness of other celestially accelerating bodies such as the Moon, planets of the solar system, and the Sun itself are synergetically balanced interattractively upon the Goddard-considered object, whereat the from-Earth-progressively-accelerated object would maintain its own cosmic orbit, though if decelerated sufficiently, it would yield to the Earth's ever more interattractive pull and thus return to the Earth's surface. The terms for this limit condition in distance and in speed are *critical proximity* and *critical speed,* respectively.

Critical speed and critical proximity constitute the independent-system-terminating acceleration that altogether demonstrates whether a celestial object is an independent system in Universe or an integral part of a larger system of energy interpatternings interknotted as matter.

I OFTEN SAW EINSTEIN ON THE STREETS of Princeton from 1951 through 1954. I and other Princeton people respected him so much that none of us ever approached him in the street. I did, however, have a fascinating indirect encounter in 1953.

Princeton's architectural department had an experimental station near the university stadium. In the years before World War I, the building had been used as the polo team's stables and dressing rooms. It was here that my students and I erected a 50-foot-diameter model of my geodesic tensegrity sphere, which I had invented several years before. It was made of ninety aluminum tubes and flexible stainless-steel cables. One day, Einstein walked over to study it. I was not there at the time, but was told by the architectural students and faculty who were there that he was extraordinarily moved by it.

The members of the Princeton community who observed Einstein's intuitive interest were so excited that they used a photograph of the tensegrity sphere on the cover of the next issue of their graduate magazine, *The Princetonian.*

None of the ninety compression struts touch one another. These nonintertouching tubular aluminum struts are held together by one comprehensive, ninety-intervaled, omniclosed-back-on-itself, spherical network of equitensed Dacron thread. If any part of this system were redundant, one of the whole-system's tension lines would not be taut. They all twang at the same pitch. If we tighten only one of the ninety intervals in the tension network, the whole system becomes equally tensed and twangs everywhere at a slightly higher pitch, indicating uniform distribution of the stressing throughout the system assembly. If we cut loose any part of the network's tension system or if we break one of the compression struts, the system does not collapse but slackens mildly, softening like a progressively deflated basketball.

Here we have a very extraordinary structure. All structural engineering today is predicated upon our Stone Age experience, in which gravity held a seemingly solid stone on top of, but not on the side of, another stone. All structuring in Universe consists of two primary forces—tension and compression. Stone masonry has high compression-resisting capability—approximately 50,000 pounds per square inch ultimate—but only 50 pounds per square inch tensile strength. Strong wood beams have an average tensile strength of 10,000 pounds per square inch, but wood fibers burn out or in time rot, not having the durability of stone. All structural engineering analysis of buildings today is predicated on what is called "compressional continuity" with only locally occurring tensional augmentation. Building construction using steel-frame and concrete reinforced with steel tension rods was not seen on our planet to any important degree until after World War I.

Compression tends to bow-bend compressional column members. Tension tends to stretch its structural members straight. Bending and buckling tends ultimately to break compression members. Straightening out tends to increase strength. Compression columns have slenderness ratios. Greek columns of stone could rise only eighteen column diameters before tending to topple over. Present-day steel columns can extend safely to only forty diameters high before tending, when loaded, to bend, buckle, and fail. Tension elements, however, have no limit ratio of diameter to length. Their invisible atom-constituted alloyed parts do not even touch one another, being held together only by virtue of the Kepler-Galileo-Newton phenomena of relative interproximity and inter-attractiveness for given masses of gravity and electromagnetism.

In 1927 I saw that the interstructuring system of Universe itself is completely different from, and magnificently superior to, structuring as thus far comprehended and employed in history by humans aboard planet Earth. Nature employs only what I call "spherical islands" of discontinuous compression and continuous tension. It is this cosmic complementation that constantly and dynamically interpositions the Earth, the Sun, and the Moon, all the stellar planetary systems, all the galaxies, and the macro and micro aspects of Universe.

Wondering whether humans are inherently barred from that level of structural design science, I note that humans did indeed invent the wire bicycle wheel. The wire wheel has a compressional atoll-rim with a hub acting as a central island of compression. The whole wire-wheel assembly takes and holds its shape only by virtue of its twelve spokes—six positively and six negatively intertensioned—and rim.

With the wire wheel humans made the historic breakthrough to discontinuous-compression, continuous-tension structures. Next, wondering whether it would be possible to produce such tensional-integrity ("tensegrity") structures in a spherically symmetrical array, I invented such a structure at Black Mountain College in 1948. Two years later, I made 3-foot, 6-foot, and 12-foot (in diameter) tensegrity spheres. Then, in 1953, I built the 50-foot-diameter tensegrity sphere that caught Einstein's attention when it was constructed at Princeton.

Universe has its radially explosive, compressional, outwardly pushing *radiation* and omniembracing, intertensing *gravity*. The total of cosmic radiation (compression) and the total of cosmic gravity (tension) comprise equal amounts of energy. Gravitation and radiation, however, operate differently. Their respective interpatternings differ. Radiation is beamable (i.e., focusable). Radiation has shadows, whereas gravity has none. Unfocusable gravity is always comprehensive; tension is always embracingly comprehensive of compression. Compression and radiation are always open-ended systems. Tension and gravity are always closed systems.

Here is a simple way of thinking about the difference between the compression-patterning and the tension-patterning of Universe. Think of a camera tripod's three legs. Since the feet of its legs are usually slippery, think of them as tending to slide apart. This happens because the three legs are fastened together only at the hinge-interlinked top end.

A force-vector is a line whose length is the product of a given system's mass and its velocity as it operates in a given known direction in respect to a known axis of angular reference. We will now assume those three sliding-apart tripod legs to be vectors of a given magnitude— that is, of equal length—joined to one another only at the top end. We then take three more tubes of the same metal and dimensions as the three camera tripod legs and fasten them together at both ends to form a closed system triangle. This base triangle tensionally (integratingly) prevents the three compressionally disintegrating legs from sliding further apart. This demonstrates that the three *gravity* vectors are integrated as a closed-system triangle, with both ends of each tensed vector interfastened with its two adjacent vectors. This closed system is in contrast to the compressional tripod vectors, which constitute an open-ended, disintegrating *radiation* system, being fastened together only at one end. The amount of energy of Universe operating as gravity is exactly equal to the amount of energy in Universe operating as radiation. However, the gravitational operating pattern of integration (tension) is always twice as effective as the disintegrative, single-ended interpatterning of the energy operating as radiation (compression) (see Fig. 3.3).

Disintegrating arrangements of radiation behaviors in Universe are always such that interpatterned gravity operates twice as effectively, which explains the integrity of eternally regenerative Universe.

Tensegrity spheres such as the Princeton 50-footer constitute a realized model of the principles governing the structural integrity of the generalized radiation and gravity field, the unified field equation which Einstein sought to express mathematically.

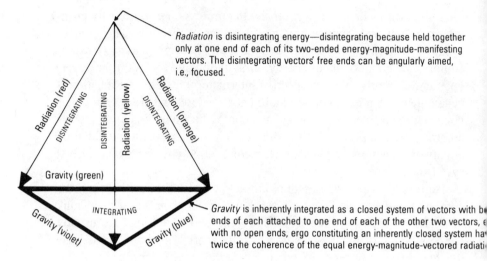

FIG. 3.3 *Gravity is inherently integrated as a closed system with no ends and ergo is an inherently closed system having twice the coherence integrity of equally energy-vectored radiation.*

In a tensegrity structure, radiation/matter is modeled by the discontinuous struts, and gravitation is modeled by the continuous network of wires unifying the structure. This model reconciles these two disparate elements into a single unified field. No other known model does so.

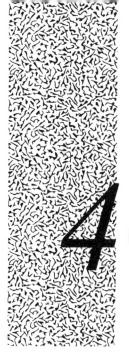

4 HISTORICAL UNDERPINNINGS

TO APPRECIATE IN PROPER PERSPECTIVE Albert Einstein's epochal contributions to humanity, it is necessary to jump backward to the dawn of civilization.

First manifest in Austronesia, then in India, then in Babylon four millennia and more ago, a loosely connected network of priest-navigators had developed a powerful epistemological, cosmological, conceptual, spherical geometry of generalized mathematical significance. This apparently was inherent in the total inventory of human experience. There is ever-increasing proof that the navigator-priests of 4000 B.C. knew by secret legend that we occupy a spherical planet around which, their verbal records reported, humans had long ago navigated. This fact was memorialized by the navigator-priests of Egypt, who designed the symbol worn on the Egyptian pharaoh's brow. This golden device was a sphere around which two snakes headed in opposite directions but came together as the planet Earth. It was a closed-system sphere. The two snakes represented the two ways these ancient sea kings had ventured forth around the world—westward with the Sun and eastward against the Sun.

These navigator-priests were alluding to what I call systemic thinking, in which, as previously explained, a system divides all Universe into all the Universe outside the system, all the Universe inside the system, and the little bit of the Universe constituting the system that does the inward-outward dividing.

Babylonian geometry is spherical—omnidirectional. The Babylonians discovered that the largest number of symmetrically identical geometric polygons into which the surface of a system can be uniformly divided is 20 equiangular triangles. The spherical polyhedron they discovered is now known by its Greek name—*icosahedron*. The Babylonians then discovered that all equiangular triangles can be subdivided by three lines drawn perpendicularly from each of their three corners to impinge upon the midpoints of their respective three opposite sides. These three perpendicular bisectors of the equiangular triangle divided the equiangular triangle into 6 right triangles, each with planar corners of 30°, 60°, and 90°. These 6 identical right triangles in each of the 20 equiangular triangles of the planar-faceted icosahedron produce a total of 120 identical right triangles per system. These 120 geometrical figures constitute the largest number of identical figures into which any cosmic system can be uniformly, angularly, and symmetrically subdivided. In the spherical icosahedron there are 60 positive and 60 negative right-angled triangles. The Babylonians also discovered and made models of the spherical icosahedron produced by fifteen great circles, all of whose 120 identical spherical triangles' three corners are 60°, 90°, and 36° each. In plane geometry the sums of the angles of all triangles are always 180°. In spherical trigonometry, used in surveying and navigation, the sums of the angles of spherical triangles are always more than 180°. The sums vary among different spherical triangles. Whatever amount over 180° the sum of the angles of spherical triangles may be is called spherical excess. Each of the 120 spherical triangles of the spherical icosahedron has a spherical excess of 6°. Thus, the total spherical excess of the icosahedron is 6° × 120, or 720°.[1]

It was from this discovery that the Babylonians evolved the "sixtiness" concept, from which we get our 60 minutes and 60 seconds of time, as well as fractions of a 360° circle with 60 minutes and seconds of angles, as being nature's optimum subdividing number. The Babylonians hoped thus to coordinate cosmic time and cosmic space calculations, but they failed in the attempt.

Such fundamental cosmic thinking in ancient Babylonia in the period

[1] Note that this number is, significantly, the precise total of degrees of all the angles of one tetrahedron.

around 1000 to 300 B.C. retrogressed as the successive world-supreme, behind-the-scenes military, financial, and religious power structures moved civilization ever west-northwestward.

Looking to discover where these early navigator-priests first came by their mathematical knowledge and systemic thinking, we must look beyond Babylonia to the very dawn of humankind in Austronesia, where we find humans born naked, helpless, and hungry but possessed of uniquely clever minds to aid in survival.

As can be most easily seen on my Dymaxion World Map (which is free of visible distortion of size or shape of any of the land areas), with propitious-for-life temperature zones indicated by color, humanity begins populating the planet Earth in the lush warm-water lagoons of the South Pacific and North Indian Ocean coral atolls.

Geographically, these people may be called Austronesians. As environmental evoluting conditions permitted, these early humans intuitively first ventured forth on rafts, riding the currents over the total Pacific, eventually inventing intentionally windward-sailing craft. Going to far-off islands and bringing back strange or magical objects, the navigator-priests helped the chieftains of the South Pacific maintain their God-ordained power. With new westward navigation capability, the Polynesians followed the life-giving Sun's seemingly ever-westward passage, which took them into the Indian Ocean and then across it to East Africa and Arabia.

Very early in the prehistory of humanity, bands of sheepskin-clothed, South China–emanating horsemen succeeded in traveling north of the Himalayas all the way to the eastern boundaries of Europe via China and the steppes of Russia. Many generations later, emerging from caves after the Ice Age, they rode southward on horseback into Greece and Asia Minor, overwhelming the people to the south in Egypt and Mesopotamia. These horse-mounted pioneers and traders extended the Far East caravan routes to the Mediterranean shores of Europe and the Levant.

The fair-haired Dorians of Greece and the horse-mounted hordes from the north who came to overwhelm early Egypt were probably descended from the Orient-sprung horsemen who had much earlier passed westward, north of the Himalayas, and apparently had survived and bleached in European caves through the entire Ice Age. Overland caravan routes were eventually established, guarded intermittently by great city-state-constructed hilltop bastions. When these castellos were attacked, the year's harvest was stashed within the walls, with the enemy left outside to starve or go away.

While vastly larger cargos can be carried on seagoing vessels than on the backs of animals or humans, the first hundred thousand years of boat development produced only open rafts, catamarans, and dugouts, which were unusable for carrying over great distances the types of goods that perish when wet. Not until stoutly keeled and ribbed ships with dry holds were developed in Southeast Asia could economically important cargos be carried. Such ever-more-ambitious shipbuilding eventually produced heavily keeled and ribbed ships that could carry valuable cargoes across the Indian Ocean. Such commerce delivered for ten thousand years its goods from the aeons-earlier-amassed wealth of the Orient to Babylon, Egypt, Persia, Mombasa, Madagascar, and the Levant.

The Arabian seacoast peoples developed shipbuilding so that ever-larger and more-profitable cargo carriers could be constructed. As we noted earlier in describing the law of similitude at length, doubling a ship's length increases its payload volume by a factor of 8 but only increases by a factor of 4 the hull surface to be built and driven through the sea. Such economies brought about the transition of land capitalism to merchant-vessel private enterprise. These sailing ships interlinked the waterfronts of Southeast Asia, the Indian subcontinent, the Arabian peninsula, and the north-northeast African coast. Great fleets followed the seasonal changes of what sailors called the "monsoon seas." Here the winds flow eastward for part of the year and westward for the rest of the year. In all of the ports along all of those coastlines, a crossbreeding waterfront people produced an ever more crossbred world sailor, identified historically with the people we know as the Phoenicians, Venetians, Vikings, Frisians, Portuguese, Celts, and some people of color—and, mythically, Sinbad the Sailor.

Far East–Near East cargo-carrying ships and their ever-evolutingly advancing designs brought supreme economic power and world trade into the hands of those who controlled the shipbuilding operations. Eventually it was these traders who became known as the Phoenicians and who navigated and traded the known world.

The fall of Troy represented the first time in which the waterborne line of supply established by the big, new seagoing vessels outlasted the dwellers within the city-state walls, who were eventually starved into surrender. The great wooden "horse," which was the Trojans' undoing, was in fact a relic of the newly massive, high-seas-going cargo and fishing ships of the new water-route masters of the ever more powerful, northwestward-spiraling flow of human civilization.

Ships constituted a mobile "environment control" allowing innately naked man to venture forth seasonally into climes too cold for previous survival. Boats could be turned keel skyward on beaches, producing the roofs of winter shelters in locales where humanity had never before lived.

The fall of Troy marked the beginning of the high-seas mastery of the large sailed and rowed ships that soon controlled the principal sea-lanes of westward world supply.

Eventually this potential wealth ignited the ambitions of Alexander the Great of Macedonian Greece, who set out to establish a new world empire based on Europe-bound commerce from the Orient via the Near East.

All this activity brought about a concentration of the most advanced knowledge of the scientist-philosophers of the Mediterranean along the Levant, first in Ionian Greece and then in Alexandria, one of the principal entrepôts of trade in the Mediterranean's westbound commerce with both North African and South European shores. Unfortified Venice demonstrated the rise to supremacy of waterborne lines of supply over overland transport. Venice's ship-supported troops overwhelmed other Italian city-states.

During the second and first centuries B.C., Italy became the westward successor of the vast military and commercial might amassed by Alexander the Great. But eventually the leaders of Rome found to their dismay that their physical might and majesty did not impress the great masses of humanity. A much more powerful and inspiring metaphysical-philosophical trending of humanity toward a belief in one God encouraged intuitive acceptance of the existence of one God of great intellect and physical power. This sole God was responsible for the design and operation of Universe, instead of, as previously thought, there being a plurality of gods, each governing special domains of Universe with appropriately unique capability. These were the household or pagan gods of the specialized slave peoples who tilled the soil and produced life support for the wealthy.

About 600 B.C. in the Orient, Gautama Buddha, an individual of royal blood, had divested himself of special privilege and attributes and become a common human. His philosophical leadership endeared him to the people, and he developed great power amongst the masses.

Six hundred years later, the westward-bound caravans brought the Buddhist philosophy from the Far East to the Near East. Its coming was symbolized by tales of camel-borne "incense and myrrh-bearing" Orientals prognosticating a Star of the Orient guiding the overland navigation of three wise men of the East. The Far East–originated Buddhist philosophic concepts and precepts thus emerged in the mid–Near East in the teachings of Christ. About six hundred years later, the same fundamental philosophy and behavioral laws once more emerged in the teachings of Muhammad.

In the Alexandrian and post-Alexandrian Near East, both the illiterate peasantry and the educated nobility, as well as the most powerful military and mercantile leaders, depended on their navigator-priests for counsel on how to please or appease the mysteriously omnipotent, omniscient God. The people were concerned about the next world, where life—in contradistinction to this world—was either an utterly idealistic reward or an eternal punishment. The priest became the authority on what a human must do to obtain the reward and avoid the punishment.

We had seen this priesthood before, for it had evolved from the early brotherhood of overland or overseas navigators, who had learned through trigonometry, astronomy, and other means to steer courses over great distances with naught but the stars and their interpositionings with self to guide them.

The navigators had the ability to persuade the noninitiates that they could get humanity from here to a predictable there at predictable times of arrival in a way utterly incomprehensible to the many. This impressive capability obviously engendered popular confidence in the priests' navigational instructions on avoiding rocks, shoals, whirlpools, and storms of this world and on safely navigating into the harbors of the desirable other world.

The navigator-mathematician-priests long, long ago realized that their power could be greatly enhanced by maintaining a general condition of secrecy, obscurity, and outright mystery regarding the origins of their knowledge and authority.

The metaphysical concept of many gods was supplanted by monotheism in the Occident at about the same time as the new post-Alexandrian, Europe-based power over world trade assumed mastery over both the overland trade routes, with their Roman road and aqueduct construction, and the sea-lanes, as the powerful Roman navy evolved from the Alexandrian ''thousand-ship'' building techniques. For the first time in history there arose a comprehensively consolidated world military and commercial power that was finding its might to be held as naught by its people, who were entirely preoccupied with the counseling of their navigator-priests. The oppressed cried out defiantly, ''We who are about to die salute you with joy in our hearts because we are bound to Heaven and you, Mr. Emperor, are bound to Hell. Let your beast come at us. The quicker we're killed, the sooner we'll reach Heaven.''

The astute priesthood was copiously informed by the people's comprehensive regular confessions, prescribed by the priest as essential to obtaining their passports to Heaven. The priests were the first to realize the military authority's inability to rule the hearts and minds of the physically conquered masses. The priests knew that they controlled the one and only popularly accredited escape from a hellish existence into a heavenly afterlife. Life in this world was a succession of misfortunes. Belief in God was necessary not only to endure the living misfortunes but also to qualify the individual for the blessings of a heavenly next world. The authority of the navigator-priest was thus at its peak.

At around the same time, for reasons still unknown to us, a retrogression in mathematical conceptioning emerges, possibly because the navigator-priests foresaw that their power would be undermined if the kings or other people caught on to too much of their calculating capability. For millennia some of their most elementary concepts would be lost: that the tetrahedron, octahedron, and icosahedron are finitely closed systems because they inherently separate the outside portion of the Universe from the inside portion (i.e., the macrocosm from the microcosm); that the triangle is the only polygon that inherently holds its shape, and thus all structure is inherently triangular; that the tetrahedron, octahedron, and icosahedron, all triangularly faceted, are nature's only three primitive structural systems.

Two millennia after Babylon we find Greek geometry robbed of systemic thinking and commencing its wrongheaded explorations with two-dimensional plane geometry. To plane geometers the world was an infinitely extended plane. The Euclideans strictly stated the rules of plane geometry, using a straightedge, a divider, and a scribing tool. They stated that the plane surface on which they scribed was that of the planet Earth. They had lost the concept of a finitely closed "system." Their thinking had been fractionated—desynergized.

Apparently an historically unrecorded conspiracy had occurred, as the priest-navigator hierarchy had, over time, secretly merged its metaphysical power and religious authority with the temporal power of the world's military and commercial masters. The conspirators had realized that the independent natural scientist-thinkers' discoveries frequently embarrassed the power structure's fortuitously contrived cosmology, which purposefully oversimplified explanations to humanity of astronomical and other scientific observations. Thus, the known world's priests, military leaders, and merchant bankers apparently agreed upon referencing all public question-answering to the priests' own convention of a very simple cosmological scheme of reality.

An earlier manifest of the power of the priesthood to do this was its capability to have Egyptian artists see and paint animals and humans only in profile (two dimensions), as an evolutionary pictorial development of much earlier Austronesian and Indonesian shadow puppets.

A second manifest of the priesthood's power was the degrading of mathematical conceptioning by society's scientific leaders, who abandoned the Babylonian's spherical mensuration and reduced the Ionian Greek Euclidean's art and science of geometry to the two-dimensional level, thus assuming the false reality of a plane-geometry world extendible laterally in all directions to infinity. This false world was timeless, weightless, temperatureless. It was a cubical coordinate system whose squares and cubes were geometrically irreconcilable with a spherical Earth and all other radiationally and gravitationally divergent-convergent, inherently nucleated, finite, spherical systems with growths and shrinkages and electromagnetic and acoustic, spherically gradient wave propagations.

The Euclidean geometers, however, felt themselves to be scientifically rigorous because they started with three tools—straightedge, dividers, and scribe (their line-scratching tool)—and could make figures only with those tools. Scientifically speaking, what they overlooked was the true nature of the surface on which they scribed.

Because the Earth is so big and humans are so relatively small, it was easy to misassume that our world surface is indeed a plane, infinitely extended in all lateral directions, and that the Earth is the center of the Universe around which the Sun, Moon, and stars revolve.

The Aristotelian-era priests told the people, including all the scholars, that the surface of the Earth on which they stood extended laterally outward from where they stood to infinity; the limits of the scribed-on world surface were indefinable and need not be included in the geometry-initiating tool inventory.

Geometry, said the Euclideans, begins with one such infinite horizontally extensive plane on which you scribe. The Euclideans apparently knew nothing of the fact that a thousand years earlier the Babylonians, informed by the legends of their around-the-Earth-sailing navigator-priests, had been manifesting expertise with omnifinite spherical-system geometry and trigonometry.

A true geometrical plane is definable only as the set of all the most economical interrelationships of three points.

A couple of millennia later, synergetic geometry has reestablished the fact that a geometrical plane can occur and can be scientifically demonstrated to physically exist only as a surface-facet triangle of any polyhedral system. The minimum such physical demonstration of four inter-edge-bonded triangles is that producing the tetrahedron. You cannot experientially demonstrate a finite nothing—much less the surface of nothing, and much less a fractional part of nothing.

It was also easy for the student of Euclid's geometry to assume that in respect to the infinitely extensive lateral plane on which we live, there may coexist an infinite number of equiinterdistanced, parallel-to-one-another planes above and below, and parallel to, plane X. This set of infinitely extendable parallel planes, together with plane X, we will call the X set of equiinterdistanced, parallel-to-one-another planes. There also exist two other sets, Y and Z, of uniformly interdistanced and parallel-to-one-another planes which are perpendicularly interaddressed to one another and which, as the combined Y-Z set, may be perpendicularly addressed to the X set of planes.

These three omniinterperpendicular XYZ (i.e., omniright-angle, interaddressed) sets will produce an infinite aggregate of to-infinity, extensive cubes. This Euclidean conception of cubically arrayed space produces what has since been known as the XYZ frame of reference. In this right-angled matrix, the vertical Y planes ran north and south and the Z planes ran east and west.

Since all the perpendiculars to the X plane are parallel to one another, they go in only two opposite directions: up and down. How far up Heaven and how far down Hell might be, was not known.

The minimum something is a system that must have both an insideness and an outsideness. A system is finite and, as stated before, inherently divides all of the Universe into all the Universe outside the system, the macrocosm; all of the Universe inside the system, the microcosm; and the remainder of the Universe, which is the closed-back-on-itself, finite system that does the macro-micro-finite dividing. All systems are finite.

The Euclideans defined a triangle as ''an area bound by a closed line of three edges and three angles'' and a square as ''an area bound by a closed line of four equal-lengthed edges and four equal angles.'' Geometrical boundaries such as those of a triangle or square identified the finiteness of only the area locally surrounded within the visual limits of the observer. The triangle scratched on the ground was misassumedly surrounded by an area that ran forever unboundedly away and was therefore undefinable. This greatly oversimplified truth made it easy for the priest effectively to misinform his listeners. The definable was only the locally boundable and thus identifiable; the finite surface of the Earth outside the scribed triangle or square was ignored. This concept greatly pleased the inherent self-interests of the landowning citizens, who might have been heard to shout, "My area—get out of here, you foreigner."

Though little specifically is known of them, the centuries-earlier Greek Pythagoreans, apparently operating from experimentally verifiable evidence, were prone to commence their mensuration with multidimensional phenomena. Long after Pythagoras, Plato's "solids" became the manifest of his multidimensional concern; of the Platonic solids, only the cube was volumetrically commensurate with Euclidean three-dimensional calculating.

In Plato's time, before the Euclidean retrogression in geometrical conceptioning, Eudoxus (c. 408–355 B.C.) initiated what is now generally considered by scientific historians to be the beginning of scientific astronomy. Eudoxus' kinematic theory of heliocentric spheres mathematically explained the complex interpatterning of any planetary system. The astronomer-mathematician Hipparchus made many corrections and improvements to Eudoxus' theories.

Philolaus, contemporary of Plato and Eudoxus, but not acknowledged by Plato, said, "The Universe is both spherical and limited. The Earth is a planet and, like the others, revolves on its axis. The Earth, however, is not necessarily at its Universe's center." Heracleides (388–310 B.C.) also established for himself that Venus and Mercury revolved around the Sun.

Aristarchus (c. third century B.C.) likewise saw all the planets to be orbiting the Sun. It was around this time that Eratosthenes measured the circumference of our Earth-sphere to within an almost negligible degree of error. In the same year, Crates developed history's first known world globe.

Historical records show that around 200 B.C. the accredited scientists of those times progressively reverted to cosmologies whose schemata contradicted much of the great inventory of experientially observed evidence with which the great Greek philosophers had so brilliantly reasoned and conjectured. Political pressure was clearly causing the scientists to abandon truth—to abandon the comprehensive and synergized facts of earlier experimental evidence.

By 200 B.C. the priesthood had accomplished its union with the military and with the latter's always discreetly hidden partners, the wealthy commerce and banking leaders, who controlled the complex economic exploitations of an amalgamated power structure. First destroying the great library at Alexandria, the power-structure-backed priesthood methodically discredited the Greek scientists' evolutionary trending toward conceiving of a solar system. The cosmology and cosmogony reverted to a flat-disk Earth surrounded in turn by a water disk that extended to infinity in the same flat plane as that of the Earth. The Sun, the Moon, and the planets revolved around this Earth disk. To complete the picture, this conspiracy soon advanced to establishing the "Holy" Roman Empire at the center of the Earth disk.

The priesthood established the divine authority of the emperor-pope of the holy empire, and in time, this authority was conceived to be conveyed to the priesthood by the disciples of the Son of God, and thereby indirectly by the Son of God, and ergo by God himself.

The cosmological model employed in explaining the experiences of life to all the people was one in which the great emperor-pope, as the supreme authority, had to be resident at the center of this flat Earth, with Sun, Moon, and stars revolving around the emperor-pope's headquarters.

To the world power structure, the solar system theory, with the Earth as one of a number of planets revolving around the Sun—and with the high probability of similar systems with planets revolving around each of the visible stars—was intolerable. The rich and powerful wished to convince the people that their living leaders were situated right at the center of Universe. This fantastic worldview is one that is still powerfully persuasive to many humans.

"Anybody can see," said the priest-strategists, "that the Sun, Moon, and all the stars revolve around us. They all rise in the morning, travel across the heavens, and descend through some part of the infinite flat plane world, probably plunging through the sea into the underworld and traveling through Hell to rise again through the sea in the morning."

It was essential to the religio-military-economic conspiracy that the people conceive of the world as flat. All perpendiculars to the same Euclidean plane became demonstrably parallel to one another. All the people, trees, and temple columns were obviously parallel to one another. All these perpendicular parallel lines went in but two directions— up and down. This directional orientation was essential to the emperor-pope's authority. Only the emperor-pope or his priests could arrange your ascendance into Heaven. Not believing in the state religion as *the* authority of God meant certain descent into Hell.

At about this time, these self-proclaimed prophets of divine authority instituted Roman numerals as the only means of expressing numbers. Thus, all mathematical calculation by ordinary people was frustrated. It became essential to the Roman power structure that nobody be able to do mathematical calculations. All calculating was monopolized by the emperor-pope's administration, giving them control of all the wealth produced by human ingenuity and labor. Roman numerals were introduced solely for scorekeeping—counting sheep. They defied use as multiplying and dividing devices and thus made all complex calculating very difficult. Calculation became the sole domain of the power structure.

It became increasingly clear that the supreme political and religious power structures of planet Earth started in Alexandrian Egypt about 250 B.C. deliberately to erase from human conceptioning the solar system discoveries of the great cosmically exploring, scientifically observing, measuring, and thinking Greeks. To rid society of every vestige of thought of the great thinkers and philosophers of former times, the power-structure conspirators destroyed the great library at Alexandria and imposed the Dark Ages view of Universe upon humanity.

All of cosmogonical and cosmological falsification dictated by the power structure resulted in frustrating the pursuit of knowledge and plunged the world into the Dark Ages. The resultant mischief and mis-conceptioning still governs human life on planet Earth.

Fortunately for humanity, an Arabian line of scientific communication from India and the Orient traveled westward via North Africa, gradually carrying the concept of the cipher, which eventually enabled scholars to develop the decimal system. This process of cumulative leftward positioning (each complete ten-finger increment or module being entered into the next leftward column as a single integer) facilitated mathematical calculation and the development of technology.

In A.D. 700, despite Rome's control over the northern shores of the Mediterranean, the Arabic numerals and concomitant system of calculation entered the Mediterranean world through the Arabian language. This westward migration of calculating capability was made possible by the simplicity of the Arabian system. As we have discussed earlier, the leftward positioning of numbers in increments of ten was made possible by use of the cipher to symbolize an empty column. The calculating capability provided by the cipher was somehow overlooked by the Mediterranean world in A.D. 700, when use of the Arabic numerals came to be allowed by the Roman church.

To the Mediterranean world the Arabic numerals had meaning and were much easier to write than the long Roman numerations. No significance, however, was seen in the cipher. You cannot see "no sheep." You cannot be hungry for "no sheep." You cannot eat "no sheep." Thus, the Mediterranean world inadvertently adopted the symbol 0 as a mere decorative device or for termination of a communication passage (i.e., as a period).

In A.D. 1200, five hundred years after it was written, al-Kwarizmi's treatise on the cipher was translated into Latin in North Africa. Two hundred years later, around 1400, knowledge of the calculating capability of the positioning of numbers was communicated across the Strait of Gibraltar from North Africa into Portugal and thence into southern Germany and northern Italy. Columbus, acquiring this knowledge of Arabic numeral calculation in Portugal, was enabled to commence thinking of a spherical Earth and performing navigational spherical trigonometry.

The Inquisition of the Roman church imprisoned, tried, and almost muted Galileo. It hoped to suppress the proliferation of any scientific knowledge that tended to imply that the Sun revolved around the Earth and that the pope-emperor and his planet were not the center of the Universe.

The church's cruel Inquisition was of no avail. Ability to calculate had been irretrievably restored to human individuals.

Despite great accomplishment during recent times, the scientists of today still live primarily in a three-dimensional Dark Ages reality, teaching their students only the misinformed *XYZ*, perpendicular and parallel, cubical and square systems of geometry and mensuration.

Despite Copernicus's embarrassment of Rome with his announcement around 1512 that the Sun, not the Earth, was the center of the solar system, the schools of today throughout the world are yet deeply immersed in Dark Ages thinking. Many of the world's leading scientists (who have known for five hundred years that the Sun does not "go down") thoughtlessly and carelessly tell their students and their own children to watch "the Sun go down" at dusk.

If you, the reader of these lines, personally use the words *up* and *down,* you, too, are as yet imprisoned in the Dark Ages. There are no parallels on the surface of the Earth; what may appear to be parallel radiates from the center of the Earth. There is no up and down in Universe; in, out, and around exist. The words *up* and *down* are relics of a time when the emperor-pope, as director of all traffic to Heaven and Hell, indicated direction with a point of his thumb.

Today's schools at every level are almost completely vitiated by the Dark Ages–imposed ignorance. Omnispecialized educational systems and the narrow professionalism they foster, together with the power structures of big money, big religion, and big politics are all still deliberately frustrating human comprehension and the possible advantage to be gained from the knowledge learned during millions of years of trial-and-error striving. In official America and Europe the criterion for success in life is making money, not making sense and not individual access to nature's own thinking and grand design.

All those around-the-world humans who saw and heard on television the landing of humans on the Moon in 1969 also heard the president of the United States of America congratulating the astronauts on "getting up" to the Moon and also heard the astronauts talking about "being up here on the Moon." At the moment they were saying that, they and the Moon were on the other side of the planet Earth from where I was viewing the broadcast. They were in fact far from up: they were in the direction of my feet.

Because of humanity's still debilitating cosmic misorientation, it was only the very reliable light-sensitive vacuum tubes, focus-locked onto the Sun and other stars and calculations progressively made by the computer, that made possible humans being safely ferried over to the Moon and back as the revolving Earth and its co-orbiting Moon together zoomed around the Sun at 60,000 miles per hour in an up-and-downless Universe.

A few years earlier, rockets aimed at the Moon by calculations of science's best mathematical minds had missed their mark by 40,000 miles or more. It was only the light-sensitive instruments' fix on the Sun and other major reference stars and a trigonometrically programmed computer that finally brought the flight path under control. Neither up nor down came into play.

In place of the words *down* and *up,* the correct words are *in* and *out*—into the Moon or into the Earth. *Out* is any direction. *In* is always directionally specific and point-to-able.

The travel directions of *in, out,* and *around* for such and such amounts of time at such and such speeds are sufficient cosmic flight data to get you to any specific location in our solar system or beyond.

Physics has found no separately demonstrable physical dimensions; no separate one-dimensional lines; no separate two-dimensional planes; no separate three-dimensional, timeless, weightless, temperatureless cubes; no straight lines or flat-out planes extending to infinity. Neither has physics demonstrated the existence of anything adamantinely solid.

Physics has found only waves of discontinuous, systemically finite energy-event constellations. There is no up or down. There is no geographically discrete wind-producing headquarters in a place called "northwest" from which the wind is said to blow—remember the cherubim shown puffing from the corners of the Dark Ages maps.

Neither winds nor columns nor spars of any great length can be linearly extended or pushed anywhere. Pushed lines curve; pulled lines tend to straighten out. Winds can be pulled (sucked) by low pressure around and about any multicornered course. When the wind is said to be blowing from the northwest, it is in fact being tensionally drawn by a low-pressure center southeast of the observer. If you think you are bravely facing into the wind, you are in fact looking in exactly the opposite direction from the causative event. Only now are the world's leading meteorologists realizing this.

Now realizing the still powerful hold of the Dark Ages on human reflex and thought, I am going once again to review my speculative working assumption of the catalytic effect on Einstein's synergetic thinking upon his learning that all radiation has the same speed.

Radiation, whether visible or invisible, is energy. All superficially disparate physical manifests of radiation, when unfettered in vacuo, have the same velocity. There are several immediate reorientations of human thinking that resulted from the discovery that all radiation (light, X rays, photons, etc.) has the same common velocity. The examples I use are my own, but in principle they illustrate concepts of Einstein.

When we look at the North Star, we are looking at a live show taking place 470 years away and ago. It has taken that much time for the light to reach us this very moment. When we look at Andromeda, we are seeing a live show taking place 2.2 million years away and ago.

These light-years-differentialed celestial displays caused Einstein to say that the observed Universe around us is "an aggregate of nonsimultaneous events." He went on to note that all those nonsimultaneously occurring, observable events are energy-radiating events of various magnitudes whose different durations overlap.

Einstein also noted that the light they radiated consists of photons and that photons are finite packages of light.

Einstein then reasoned that an aggregate of finites is finite and, though the whole of the Universe cannot be witnessed simultaneously, inasmuch as it is an aggregate of finites, despite its nonsimultaneous viewability, it must be finite. This, of course, was an entirely new way of thinking about Universe.

From all the conceptioning, considering, and conceiving of Einstein, I concluded that what he had discovered was what I refer to as "scenario Universe," an endlessly evolving complex of dissimilar filmstrips, in contradistinction to the exclusive "single-frame" picture of Universe adopted by classical science. A scenario is an aggregate of overlappingly introduced episodes, characters, themes, and only locally included births, lives, deaths, and other events.

Though he did not express it in this way, Einstein introduced to human thinking the Dark Ages–dispelling concept of an omninonsimultaneous, eternally regenerative, only overlappingly episoded scenario Universe with all its concomitant, only locally occurring beginnings and endings.

This concept altogether superseded the Dark Ages concepts of Newtonian and classical science: single-frame, instantaneous, exclusively three-dimensionally structured, everywhere-the-same-time.

Finite and *infinite* were commonly accepted phenomena in the Dark Ages view of reality. Einstein eliminated the perception introduced by Euclid and perpetuated by Newton that an infinity of straight lines or perfectly flat planes could possibly exist. Einstein brought to the scene a new way of thinking about the experimentally derived scientific information of the existence of the Brownian movement and the discovery of the photon and blackbody radiation. He saw the Universe as an aggregate, finite but nonsimultaneously (nonunitarily) witnessable.

With vision obscured by the Dark Ages fog, we would, with little hesitation, proclaim, "All that is simultaneously conceptual is finite; that which is not is infinite." Einstein taught us to think (1) finite but nonunitarily conceptual, and (2) unitarily conceptual and definable. Universe is a nonsimultaneous and everywhere-always-closing-back-on-itself system of lesser systems of complexedly overlapping or interweaving episodes.

Each episode has its own finite beginnings and endings similar to nonsimultaneously intertwined, separate hemp fibers progressively twisted together into threads, the individually beginning and ending threads twisted into strands, and those individually beginning and ending strands twisted together into rope, whose individual beginning durations of existence and endings overlap the existence of myriads of other individually beginning, enduring, and ending ropes, which complex of individuals eventually and nonsimultaneously separate and disintegrate into dust, topsoil, atoms, molecules, stars, whose sidereal radiation is photosynthetically integrated biologically to nonsimultaneously produce, for instance, hemp fibers to be harvested and twisted again into threads, and so on.

Nothing is lost. This principle is the driving force of eternally regenerative scenario Universe and shall outlast the Dark Ages and any future misconceived episodes, however and by whomever wrought.

5 TAKING INVENTORY

BEFORE THE ASCENDANCY OF THE BRITISH EMPIRE, all previous empires of history, such as those of Alexander the Great, of the Romans, and of Genghis Khan, were flat-world empires. No one knew what went on beyond the map's borders. The British Empire, securely established in 1805 with the great sea battle at Trafalgar, was the first empire in history on which "the Sun never set." It was a spherical-world empire—the result of two hundred years of daring conquest, scientific exploration, and economic treaties. It was through the mechanism of the British East India Company that, for the first time in history, a harvest of economic, scientific, and social information from around the spherical Earth was collected and digested. Thomas Malthus, when he became professor of political economy at the East India Company College, realized that he was the first human being in all history to have the vital statistics of humanity directly collected from all around a closed-system spherical planet, as distinguished from an open system with its only locally significant economic data.

Thomas Malthus proclaimed in 1803 that the global data showed humanity's population to be increasing at a geometrical rate while its life-support productivity was increasing only at an arithmetical rate.

Based on this, he concluded something to this effect: Quite clearly the majority of humans are destined to live out their years in great want, pain, and suffering. Pray all you want, it will do you no good. That's all there is. Planet Earth has been scientifically established to be a closed system.

Although Malthus's assumption has long been assumed by economists to be a generalized scientific law—i.e., that an inherent fundamental inadequacy of life support exists on our planet—I saw as early as 1917 that technology provided an unexpected and adequate counter to his assumption and its later incarnations under the general rubric "limits-to-growth theories."

In 1859, Charles Darwin promulgated his theory of evolution, explaining his belief in the survival of only the fittest species and of the fittest individuals within those species. He later protested that he never meant his theory to have any economic significance. His contemporary Karl Marx felt that Darwin's theory of evolution clearly governed socioeconomics. While Marx did not specifically say this, his written thoughts make it eminently clear that he accepted the findings of both Malthus and Darwin. To Marx, the worker was quite clearly the fittest to survive because he knew how to use the tools and how to make all the products. The worker knew how to nurture the seed and the lamb. To Marx, the wealthy people were parasites. They did not agree, thinking, "We're on top of the heap, and Darwin's 'survival of the fittest' explains why we're on top of the heap. We're fittest. The worker is very dull, very locally preoccupied. What humanity needs is imagination, very big thinking and venturing, and a lot of courage and initiative to make the closed-system world work."

It was only a century or so ago that there occurred the fundamental ideological dichotomy in human political and economic affairs—the Communists versus the capitalists, who later preferred the appellation "private enterprise." Being known as private enterprisers rather than as capitalists suggests daringly brilliant risk taking on behalf of humanity, which warrants capitalism using its power to gain benevolent tax and subsidy advantages not made available to the public in general. The fact is that today capitalism takes the least risk of all social functions. Capitalism's prime interest is self-interest, that is, further government commitment to armaments expenditures.

It is very important to recognize that 99 percent of the people now ruled by Communism and most of those now controlled by capitalism did not elect so to be classified as Communists or capitalists. The great masses involved dreamed that they were doing what they wanted to be doing—i.e., living in a democracy. Both Communist and capitalist leaders have assumed dictatorial power to be essential to their respective successes and are ever reconnoitering to impose their ideology's viewpoint on people.

Returning to Malthus, there was 99 percent illiteracy around the world at the time he was working on his theory. His inventory of facts was in effect a wealth of highly classified information belonging exclusively to those ambitious to run the world and reap its riches. Malthus's discoveries and conclusions remained popularly unknown through the first half of the nineteenth century. His findings were of interest only to those interested in winning control of the world's wealth away from its England-based masters, since and because of Malthus's pronouncement of a fundamental inadequacy of human life support on our planet. Each of the respective ideologists said then and still say, "You may not like our system, but we're convinced we have the fairest, most logical, most ingenious way of coping with lethal inadequacy of life support on our planet. But because there are those who disagree on how to cope, it can only be resolved by the trial-of-arms which system is fittest to survive."

The foregoing explains why the Soviet Union and the United States for over four decades have spent trillions of dollars and trillions of rubles to buy the highest capability of science to discover, develop, produce, and stockpile the means to kill ever more people at ever-greater distances in ever-shorter time.

When I was born in 1895, popular reality consisted of everything that could be touched, smelt, tasted, heard, and seen with the human senses. When I was young, a new era was opening.

I was born the year X rays were discovered, the year Marconi first used the "wireless." When I was two, electrons were first identified; it did not make the news. Nobody knew that electrons would eventually have socioeconomic significance. We were entering an age when, as today, 99.999 percent of the technological reality affecting all our lives is nondirectly contactible and apprehensible by the human senses.

As already mentioned, all structuring consists of tension and compression. Historically speaking, stone and masonry had a compression-resisting strength of 50,000 pounds per square inch, in contrast to a tensile strength of only 50 pounds per square inch. The strongest available wood had an average tensile strength of 10,000 pounds per square inch.

At the time of my birth, metallurgy was developing the electrolytic refining and production of aluminum—a metal that is much lighter than steel but is not nearly as strong. Aluminum had theretofore been so difficult to produce that Napoleon had aluminum dining plates that ranked with gold plates in cost.

Suddenly, we began to develop metallic alloys of greatly increased but invisible strength. Our first mild steel production in 1851 had both tensile strength and compression-resisting strength of 50,000 pounds per square inch. In 1883 W. A. Roebling used high-carbon alloyed steel in his Brooklyn Bridge; it had a tensile strength of 70,000 pounds per square inch. In World War I—my coming-of-age era—industry developed chrome-molybdenum aircraft steel with a tensile strength of 110,000 pounds per square inch. This was more than twice the tensile strength of 1851 mild steel, yet weighed no more per unit volume than the mild steel.

In World War II we had chrome-nickel (rustless) steel with a tensile strength of 350,000 pounds per square inch of cross section. Now we have in practical use carbon fiber with a tensile strength of 600,000 pounds per square inch and with the same weight per cubic inch as the mild steel of 1851.

No one can see the differences because they are invisible. Society pays no little attention to anything invisible. Up to the time of World War I, when steel steamships replaced wooden sailing ships, everybody thought of ship sizes only in terms of Archimedean displacement (i.e., their tonnage). All the old men-o'-war were identified by the ship's tonnage and the number of ships in the armada.

There was a popular working assumption that "you can't lift yourself by your bootstraps." It was assumed that every function has a given (constant) weight and work involvement. Even today this is the economists' working assumption. Economists differentiate only between aluminum and steel, not among various alloys.

Because of the appearance of new alloys with their invisible increase in tensile performance per pound, we made a startling realization during World War I. We could defeat an enemy ship of a size equal to our own, of the same tonnage, with the same number of guns of the same caliber— everything virtually the same—if we had one all-important advantage. If our ship's biggest guns, the same size and weight as theirs, were made of steel with twice the tensile strength per pound of theirs, our guns would be able to shoot accurately at a range perhaps one thousand yards greater than theirs. Firing at them as they first came within our range, we would be able to sink an enemy ship before it even got close enough to fire at us. Such information was "secret" (i.e., critical) information.

I saw that all the most highly classified information during World War I concerned the *invisible reality* of the emergent technological revolution of continually doing more with less. Nobody could see it. Because society could not see it, such secrets were readily kept. Nobody talked about an invisible technological revolution taking place.

Because society could not see it, society did not know about it.

There are as yet no economics books—or chapters or even sentences in such books—about doing more work with the same weight of material, ergs of energy, and seconds of time or about doing ever more with ever-less resource investments per function accomplished. The one great generalized law of all economics is the fundamental inadequacy of life support on our planet.

Evolution's provision of an escape hatch from the otherwise ever more swiftly and invisibly developing consequences of the Dark Ages' haze-over became compounded with the invisible evolution's perils. Ultimately most lethal are the cosmological, academic, and everyday socioeconomic misorientations of all humanity by the insidious metaphysical influence of the Dark Ages, misassumed to have terminated long ago. These misorientations have been welded into human affairs as accepted "legal and academic" precedents and customs manifest in the world's successively dominant socioeconomic and militarily supported power structures.

To acquire essential insights regarding the strategic role of Einstein's conceptual breakthroughs in the realization of humans' potential emergence from the Dark Ages, it is necessary to comprehend realistically the part being played by the invisible structuring of metallic alloys. This is only elucidatable by Newton's law of mass interattraction and other, less well known mathematical laws.

Many scientists will not seriously accept nonmathematically expressed explanations. Because I am hopeful that some responsible scientists and engineers will comprehend the gargantuan economic significance of ever more effective performance with ever-less investment of resources and their altogether combined interfunctioning transpiring in the invisible ranges of technological evolution, I have included a mathematical elucidation of alloying as well as a verbal explanation.

Being a technologist and U.S. Navy officer of the line in World War I, I realized back in 1917 that the possibility of doing progressively ever more with ever less might mean that at some not too distant date we might attain such a magnitude of accomplishing more work with so much less resources that we would be able to take care of all humanity at an unprecedently high standard of living.

Technological invalidation of Malthus's assumption of a fundamental inadequacy of life support on our planet became my most important goal. Of course, Malthus's reading of his data was correct for his time. It was not a generalized law, however, as the economists assumed it to be. It was only a temporary condition, similar to what I saw as the situation with fossil fuels on this planet. I saw fossil fuels as a very precious resource that had taken millions of years to produce and could serve only as a temporary battery to fuel industrial growth on the planet for a relatively short period, until technology could advance to the level where all energy would come from renewable and solar sources.

I became very excited by the challenge.

Reviewing briefly my own history and its relationship to the swift evoluting changes in vital criteria, I came out of the navy and entered the building world.

By 1927, I was penniless and in abject dismay. I was certain that I would never be able to succeed financially in the competitive survival game of the peacetime business world. On the point of suicide, I determined that I had a unique set of experiences that were not mine to discard and might, given the right circumstances, have some incremental effect on the future course of humanity. To think of one individual, infinitesimal in importance in relation to human cosmic evolution, having a role in that evolution may seem to be a product of ego, megalomania, or exaggerated importance, but on that fateful day, I concluded that this relationship of the minute individual in respect to the whole is nonetheless the only possible common direct experience of each and every human being. All else is hearsay.

In order for you to understand how fortunate I have been to have had the life experiences I have had, you must get a sense of the crisis in which I found myself in 1927. I reasoned, "Since I'm really a throwaway, if, instead of committing suicide, I use my entire experience and knowledge inventory in an experiment of only working for all humans rather than one human, that commitment might validate my survival."

In 1927, when I was thirty-two, the American Institute of Architects (AIA) published an article about a single-family dwelling that they felt to be ideal under the most technologically and economically advanced circumstances of the time.

To appreciate the magnitude of 1920s improvements incorporated into that "ideal" 1927 AIA single-family dwelling, we must realize that before World War I we had sawed out blocks of pond, lake, or river ice to fill our home iceboxes. Also in our most opulent households we had coal-fired furnaces requiring coal-shovel stoking. This AIA ideal 1927 house had an electric icebox and a self-tending oil-burning furnace. Everything was "right up to the minute."

I analyzed that house as described by the AIA. I calculated its total floor area and total volume of enclosed space. I listed and work-rated all its technical facilities and characteristics. Counting all its windows and their compass-orientation, I calculated the number of lumens of sunlight entering the house. I then calculated the total weight of the AIA house, including its incoming water pipes, sewer lines, and wires. That 1927 AIA ideal single-family dwelling weighed a total of 150 tons.

Then, using the most advanced aircraft technology of the time—aircraft aluminum alloys had just been developed—I calculated the total weight of a single-family environment-control and human-life-serving machine I had designed with the same cubic footage, the same floor area, and the same technological performance capabilities. I estimated that my autonomous dwelling machine would weigh only 3 tons, as against 150 tons for the AIA's conventional-building-technology single-family house—i.e., only 2 percent of the weight of the comparable conventional building technology. That was in 1927. My Dymaxion House did not resemble the conventional AIA architecture. It had its own aeronautical look about it.

In 1945, when the interim alloy research had been completed, I built two full-scale prototypes of the Dymaxion House for the U.S. Air Force at Beech Aircraft's shops in Wichita, Kansas. These prefabricated, air-deliverable dwelling machines weighed in at exactly 3 tons, the weight I had predicted eighteen years earlier. This reaffirmed my confidence in both my understanding of design science capability and my speculative analysis.

There is an ultimate technological fallout from military production's instrument and tool development into the furnishings and appliances of the home front, such as the already mentioned refrigerator. But often the transition takes a generation or more. Mechanical refrigeration appeared in the navy twenty years before World War I and thirty-eight years before the electric fridge of the AIA house. In 1927, I posited that if we applied the most advanced aircraft and naval production capabilities *directly* to the home front, we might be able to greatly advance the realization of a livingry advantage for all humanity, eventually taking care of all humanity's physical comfort needs. I saw this as a means of shifting humanity from a failure strategy to a success strategy. I sought its implementation in all my inventions.

That is how I entered upon this fifty-five-year-long project. I could find nobody else even mildly interested in undertaking these experimental developments. I kept track of, and plotted, all the curves of rates of increase of tensile strengths in all the different kinds of metal alloys. I also started in 1927 to keep track of the increase in automotive horsepower in relation to engine weight and gallons of fuel expended.

I foresaw the ultimate development of a large plastics industry producing materials similar to our fingernails that would be opaque, translucent, or lucent and as relatively unbreakable as poker chips and fountain pen barrels, which in 1927 were among the only plastic products. At that time there were no plastic products larger than celluloid dolls. Anticipating products as large as our present-day seventy-foot-long yachts of reinforced fiberglass hulls, I predicted large, strong, and lightweight all-weather plastic reinforced by high-tensile-strength steel rods.

All of my fifty-year anticipatory planning was predicated on the up-to-then rates of increase in performance capabilities. Keeping careful track of many performance curves enabled me to make very powerful prognostications.

My integrated performance curves showed that the rates of actual increase in our ability to do so much more with so much less for so many more people made it realistic to assume that we might be able to take care of everybody at an ever-higher standard of living and do so within the twentieth century—at the slowest rate of improvement, by the year 2000; at the fastest, by 1990. There were, for instance, the curves for the per capita use of copper in the United States and in the rest of the world. There were two trends: an ever-decreasing per capita use in the United States and an ever-increasing amount for each world human. In 1936, U.S. humans had 125 pounds per capita and world humans only 15 pounds. The curves of decreasing pounds per U.S. human and increasing pounds per world human come level with one another in 1996.

In 1927 it was possible to calculate that it would take about half a century to get to a visible-to-others realization that we were indeed approaching that condition of universal technologically achieved abundance. In 1938, in *Nine Chains to the Moon,* I published some of my charts of these calculations, which means they can be reviewed today. I also published them in *Fortune*'s tenth-anniversary issue in February 1940.

The critical path for the Apollo Project's ultimate 1969 successful ferrying of humans over to the Moon and back consisted of a list of the million-plus tasks that were going to have to be done—that had never been done before—as well as a list of all the essential things that we had already proven could be done and that must now be put to use. This schedule had to be satisfied before the blast-off of that ultimately successful Moon voyage.

Using a first-things-first strategy, the critical path of the Apollo Project had to sort out and arrange the order of tasks to be accomplished for this unprecedentedly complex and massive project. With critical-path planning, each task has a precise order of subtasks to be accomplished and a schedule for each. The critical path provides in advance a master schedule of dates by which the longest-to-accomplish tasks must be initiated, with subdates for all necessary and related tasks.

In 1927 I foresaw a fifty-year critical path necessary to prove that Malthus's conclusions were limited to a special early-nineteenth-century case. I sought to prove this by demonstrating the logic of consciously implementing a high standard of living for all humanity by employing the invisible-reality technological revolution in producing livingry artifacts. My ultimate objective was to convert the most-advanced technology from producing killingry (armaments) to producing high-tech livingry.

In the widely published and discussed 1974 Club of Rome report about the "limits of growth," the authors considered the world's mines to be the only source of metals. They found that humanity had almost exhausted these mines.

So ignorant are our economists that there was no one on the Club of Rome's economics computer team at the Massachusetts Institute of Technology who knew that 70 percent of our steel comes from recirculating scrap metal or that 80 percent of our copper comes from recirculating scrap.

We have reached the point where no more mining need be done. In my tracking of resource curves, I discovered that the average of all metals recirculates every twenty-two and a half years. Some metals come out of their functional-use state very quickly, say in five years, while others come to be recycled every fifty years. Each time they come around again, we have gained so much more know-how and can do so much more for so many more people with so much less in the way of physical resources per function that ultimately we need not mine any more.

Long ago I saw that we could take the metals that are in all of our weaponry, melt them down, and implement them directly for livingry. Based on my logistical engineering experience—having had over two hundred thousand of my geodesic domes installed around the world in the most formidable arctic, antarctic, and equatorial environmental conditions—I see that it is now highly feasible to institute a millennial ten-year design revolution that could take care of all humanity at a much higher standard of living than anybody has ever known and could do so on a sustainable basis. During those ten years, we could also phase out forever all further use of fossil fuels and atomic energy. We can live entirely on our energy income from the Sun.

If you make such a statement publicly, you are sure to get rigorously checked. My contention has been checked by many specialists, none of whom, to my knowledge, has found me in error. Of the five billion human beings on our planet, possibly a million now know what I have discovered and that I am correct in my contention—that we presently have the technological option to establish five billion billionaires on our planet. I saw that humanity, largely unaware of its potential, might not exercise its options in time. A sense of urgency fueled my invention-implementation strategies, my writing, and my speaking engagements.

Though we humans are here in Universe to use our minds to discover principles and to employ them objectively, I find that today muscle, cunning, brains, fear, and selfishness are in control of human affairs—not mind. If mind were in control, or comes into control in time, we would certainly exercise our option to have everybody in ascendancy and come to a new kind of operating relationship with Universe.

6 COSMIC CONCEPTIONING

PRIOR TO THE TWENTIETH CENTURY, great scientific discoverers were prone to be comprehensivists rather than specialists. They identified themselves as "natural philosophers." Less scientifically informed leaders, who tended to integrate their total experiences into explanations of universal beginnings and endings and governance, became religionists. Navigators, as people who learned to steer by the stars and who became expert in how to get from here to there, became the guides to the next world.

It was in this way that specialization induced by prehistoric circumstance brought humanity eventually to the brink of chaos and utter destruction at the very dawn of Einstein's Universe. Humanity is now maintaining an unstable collection of local holding patterns, awaiting a physical or metaphysical integrity to give structure to the future and to show the way out of the darkness. The twentieth century's leap into a realm with a million times greater range of reality, produced by the sudden visibility and employability of the total electromagnetic spectrum, has brought humans to the edge of self-extinction for lack of adequate guiding forces. Big business and big religion's inclination for moneymaking and power has served only to foster the continuance of a millennium of isolation, inhumanity, misinformation, and ignorance.

We now have available to each of us the comprehensive information that can lead us out of the Dark Ages, which continue to hold us down with physical and moral barriers to the free flow of the information and materials that would spontaneously liberate us. The old structures were prejudicial human physical-power structures. The adamantine new structure is metaphysical, pristine, eternal, a generalized system of pure principle.

The experimentally founded mathematics that I call synergetics will disclose the geometry that we ought to be teaching our children. Synergetic geometry is the earliest systemization of the emerging information about nature's own most-economical coordinate system and the universal design principles that govern it.

All seven wonders of the ancient world were physical. A new set of seven wonders has acquired prominence with human entry into the twentieth-century realm of metaphysical reality. A list of these metaphysical wonders, some of which predate our current era, would have to include the following:

1. The invention of the cipher and concomitant positioning of numbers
2. The algebra
3. The amazingly accomplished Keplerian, Galilean, and Newtonian evolved mathematical laws of gravity and variable cosmic coherence
4. The Einstein cosmic radiation; Roemer's discovery that light has speed, and his accurate estimate of the uniform speed of all radiation, further amplified by Millikan and Einstein; Einstein's equation $E = mc^2$
5. Avogadro's law, stating that under identical conditions of heat and pressure, all gases will disclose the same number of molecules per unit volume
6. Euler's topology and Gibbs's phase rule
7. Synergetic geometry and tensegrity geodesics—vectorial coordinate system of nature—including the Einstein-initiated conceptioning, discovery, and proof of an eternally regenerative, nonsimultaneously episoded scenario Universe in which all local events are only omni-tensegrity cohered, pulsatively convergent and divergent.

Many Ph.D.-bearing mathematicians busy themselves with nonexistent objects—for example, quasitopological "surfaces" of nothing—pretending that they exist. They intensively study other fantastic phenomena: physically nondemonstrable "things" with one, two, or three dimensions, which supposed objects are ageless, weightless, colorless and temperatureless, with no inside distinguishable from outside.

They somehow base their theories on these nonexistent, nongeometrical nonentities. For instance, all geometricians, both old-fashioned and post-Euclidean, assume that a plurality of lines can go through the same point at the same time.

What cannot be experimentally proven is called *axiomatic* by geometricians and by mathematicians in general. *Axiomatic* means to them "obvious" or "it has always been taken for granted to be thus and so."

Synergetics, on the other hand, deals only with experientially demonstrable phenomena.

Specifically because no two events can transit the same point at the same time—we come to have radiation interference, which, when it reflects back to our optical system, provides human sight or, as with radar, bounces back invisibly to inform us of remote macro-otherness bodies (see Fig. 6.1). In such a manner, the electron and field-emission microscopes provide us with true microcosmic photographs of the atom.

A conceivable otherness requires a surface. Light bouncing off that surface provides the observer with optical information acknowledging its presence for relay to the brain. We cannot have a surface enclosing nothing. A surface is an outside, which inherently requires an inside. To produce an experiential model with an insideness and outsideness requires four vertexes; that is, the model must be at minimum a tetrahedron. Such a division of insideness and outsideness constitutes a system. Anything less is inconceivable.

The mathematician's purely imaginative points, lines, and planes are nonexperienceable. They cannot be modeled, having no thickness, no breadth, and ergo neither insideness nor outsideness. All imaging derives from experience. Conceptually imaginable point, line, and plane experiences are systemic; that is, they have insideness, outsideness, and angular constancy independent of size.

Size is always special-case realizability. The mathematician's undemonstrable assumption that three points define a plane of no thickness—no radial depth—is therefore subsystem, unthinkable, not operationally evidencible, unimaginable and ergo unemployable as a constituent of a proof.

Contrary to conventional mathematical dogma, three points do not define a nonexistent and ergo nondemonstrable, no-thickness plane, nor do they define an altitudeless triangle, because there can be naught to do the defining systematically. No-thickness is neither experimentally

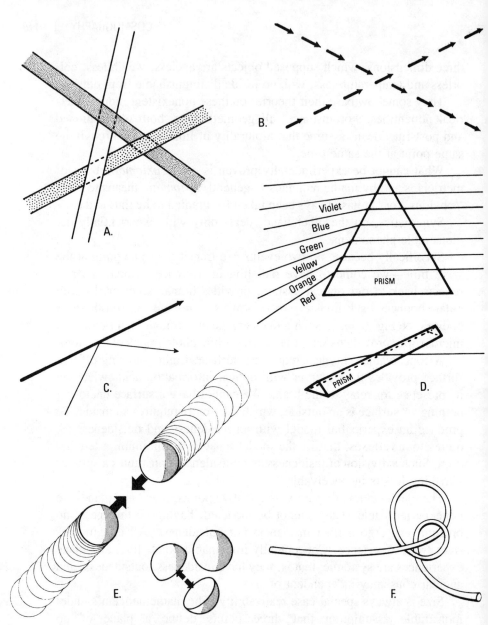

FIG. 6.1 *Interference phenomena: lines cannot go through the same point at the same time.* No two actions can go through the same point at the same time. The consequence of this can be pictured as follows:

A. Tangential avoidance (as with knitting needles)
B. Modulated noninterference
C. Reflection
D. Refraction
E. Smash-up
F. The minimum knot or critical proximity interference pattern

evincible nor conceptually feasible. System is conceptual independent of size.

Recently I was asked by a publisher to comment on the writer Annie Dillard's book *Teaching a Stone to Talk*. It got me to thinking about how I do not have any friends who can tell me so much with so few words as do the stones. In their own way, they are eloquent.

How convenient are stones for throwing into the water to watch once again the perfect circular waves concentrically emanating from even the most carelessly tossed-in, highly asymmetric stone.

To demonstrate a unit something, all we need is a single stone.

Three quarters of a century ago, my brother Wolcott and I spent day after day exploring the fascinating beaches of our Bear Island wilderness home on Penobscot Bay, Maine. I frequently reminded Wolcott of his inability to find a throwable-with-one-arm stone of any given shape that I could not make skip gracefully atop the water surface.

Taking each of his successive "challenger" stones, I would first roll it around between my two hands and toss it between them. Then I would toss it in my throwing hand, confidently determining its center of gravity and natural axis of spin. Next I would observe which of the poles of the stone's spin-axis was the flattest—most like a boomerang's undersurface.

Holding the flattest pole of the stone toward the ground, and with the index finger of my right hand curled around the stone's spin girth, I would go to the water's edge. There, half crouching, with my left foot toward the water, I would bend my throwing arm as far backward as was comfortable. Using all my strength, I would swing my arm parallel to the water's surface, just high enough above the beach to avoid touching it. I would throw the stone horizontally, inches above the still water, simultaneously imparting an accelerated spin with my elliptically curved index finger, aided by a final, jai-alai-technique wrist whip. The stone would accelerate into a precessional gyration, its flat underside spinning like a discus. The challenger stone developed a 90°, precessionally repellent force which, combined with its predominant horizontal acceleration, produced a delicate succession of concentrically circled, skim-skip, skim-skip touchdown and run-out spots.

In all my testing by Wolcott, no stone ever failed to produce that multi-skip-along path. A natural athlete, excellent engineer, and champion sailor and celestial navigator, Wolcott did not concede excellence to me in any other department than stone-skipping on water.

Stone-skipping is not an Olympic Games event, but it would make a spectacular one, requiring slow-motion television replays to verify distance and number of touchdowns.

Because rounded stones of different sizes interroll one upon another, like ball bearings of differing radii, beaches of surf-smoothed stones are difficult to walk on. They allow our feet to sink deeply into them. To produce firm roadways, stones are crushed into sharp-edged pieces, which pack ever more tightly and fixedly together.

One way to get started understanding what stones are saying is to walk over such a path or roadway made of stones that have recently been crushed into smaller pieces.

Picking stones at random and inspecting them carefully, you will soon discover that no matter how many times they are broken into smaller stones, none are ever produced with fewer than four corners or with fewer than three faces around each corner or with fewer than three edges around each face. This mathematical limit condition is descriptive of a tetrahedron. In a regular tetrahedron, all the angles are the same. You will most frequently encounter stones with an overall asymmetric form—that of an irregular tetrahedron. You are learning that nature has mathematically elegant pattern aspects that are only superficially hidden.

Stones are *always* polyhedra (many-sided) even when they appear to be polished spheroids (see Fig. 6.2). Looking through a lens of sufficient magnifying power will always reveal many mini–mountain peaks, sharp ridges, and angular plateaus. There are no perfect spheres, only polyhedra with many, many sides.

In an epochal breakthrough for both mathematics and humanity, the great eighteenth-century Swiss mathematician Leonard Euler discovered certain unique geometrical patterning rules that were later gathered together under the general rubric "topology."

He demonstrated that all visual picturing experiences are resolvable into only three unique aspects: (1) lines, (2) crossings of lines (also called points, fixes, vertexes, or corners), and (3) areas delimited by lines (also called faces or windows).

Euler further demonstrated a universal law that the number of vertexes (V) of all polyhedra plus the number of faces (F) will always equal the number of edges (E) of that polyhedron plus the number 2. Euler's formula is written $V + F = E + 2$.

To elucidate Euler further, I shall next reiterate in detail my own (not Euler's or anyone else's) unique *system* concept—unique in that it differs greatly from Ludwig von Bertalanffy's General System Theory and its many derivatives.

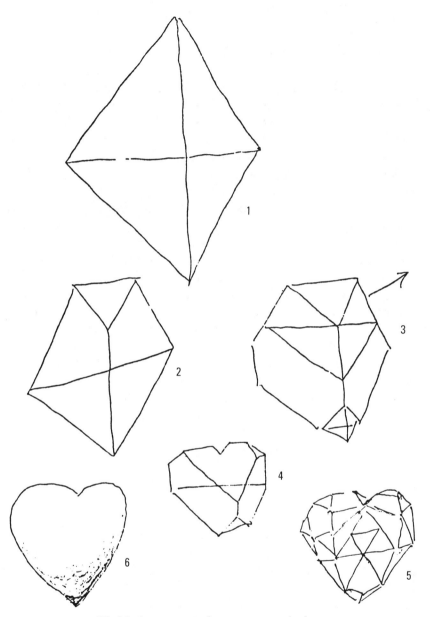

FIG. 6.2 *A stone transforms to a tetrahedron.*

System

A system is the simplest physical or metaphysical experience we humans can have. A system must always have insideness and outsideness. Recognition of a system begins with the initial discovery of either self or otherness. We recall life begins with awareness of otherness: no co-occurrent otherness, no awareness. If there is no insideness and outsideness, there is no otherness and ergo neither life nor thought.

As we have seen, systems always divide all Universe into three principal parts: the system itself; all Universe outside the system (the macrocosm); and all Universe inside the system (the microcosm).

More incisively, the foregoing three-way division can be expanded into five zones. All Universe outside the system considered is divided into (1) the clearly irrelevant macrocosm zone and (2) the twilight macrocosm zone of tantalizingly possible relevance. The next zone is (3) the system itself; clearly relevant and tuned-in, it convergently-divergently divides all Universe into its macro-outsideness and its micro-insideness irrelevancies. The microcosmic insideness is divided into (4) the twilight microcosm zone of tantalizingly possible relevance and (5) the clearly irrelevant microcosm zone.

In synergetic geometry we are able to consider the geometry of thought systems.

Thought systems encompass macro and micro twilight zones of contiguously recallable information that is intuitively considerable as being of possible relevance or even as being of significant relevance. The difference between geniuses and nongeniuses is that in addition to attending to the clearly relevant tuned-in system, the genius also pays intuitive attention to tantalizing, could-be-relevant zones of information.

All children are born geniuses, but are swiftly "degeniused" by their elders' harsh or dull dismissal of the child's intuitive sense of what could be relevant. Children spontaneously weigh all information from their immediate experience and try to relate it to other experiences of some time before. The incipient geniuses must somehow weather, year after year, the barrage of admonitions to ignore what they spontaneously think, instead only paying attention to what others think and are trying to teach.

Human mind inherently seeks comprehension of the topological interrelationships of all experiences. Geniuses discover, speak out on, and mathematically formulate the generalized principles they find underlying all experience.

A system divides all Universe, convergently and divergently separating all the outwardness from all the inwardness and from the system itself, which does the dividing. A system is unthinkaboutable. It considers all experience-generated information, spontaneously tuned-in, as relevant, dismissing all experience considerations that are too large and too low in frequency to alter in any way the clearly tuned-in conceptioning's magnitude of any one system's significance-assessing and also dismissing spontaneously all experience-considerations that are too small and too high in frequency to be of discernible significance at the tuned-in magnitude of the considered system's wavelengths and frequencies (Fig. 6.3).

In synergetics, the always and only experientially based geometry of

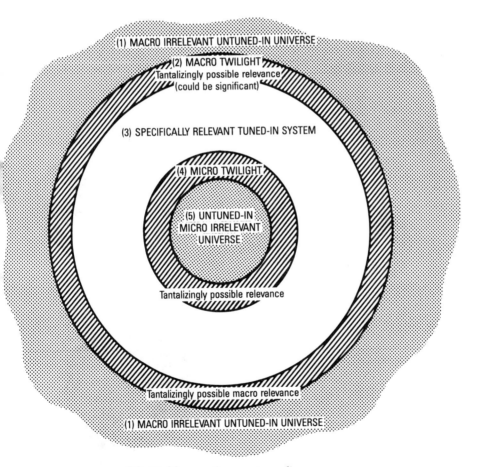

FIG. 6.3 *Macro-micro systems diagram.*

conceptualizing and thinking, I discover first that all experienceable somethings—be they apples, cows, thoughts, clouds—are *systems*.

The minimum something in Universe is a system (Fig. 6.4). There are no parts (or elements) independent of systems. A system always divides all Universe into these ten intercomplementary but distinctly different component categories:

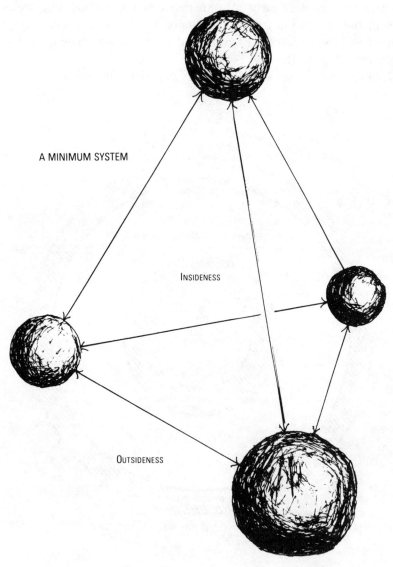

A MINIMUM SYSTEM

INSIDENESS

OUTSIDENESS

FIG. 6.4 *The minimum system.*

1. All the tuned-in Universe outside the system; the relevant macrocosm or macroenvironment outside the system
2. All the tuned-in Universe inside the system; the relevant microcosm or microenvironment within the system
3. The polyhedral constellation of Universe events defining the system itself, which divides the macrocosm from the microcosm
4. All the at-present non-tuned-in, irrelevant macroenvironment of the system
5. All the at-present non-tuned-in, irrelevant microenvironment of the system
6. All the at-present macro-Universe, large-wave, low-frequency, tuned-out (not tuned-in) programs irrelevant to the for-the-moment considered—tuned-in, felt, or thought about—system
7. All the at-present micro-Universe, short-wavelength, high-frequency programs irrelevant to the for-the-moment, tuned-in, felt, or thoughtfully considered system
8. All the recallable systems of past experience that can in no way be altered
9. All the as-yet-not-happened thinkaboutable systems of experiencing, many of which are subject to design by the individual
10. All the happening-right-now experience events, some of which are unalterable by the individual and some of which are designedly controllable by the system-concerned individual

When scientists say that they are seeking to establish the parameters of a problem, they are in fact seeking to establish all the macro- and microrelevant aspects of the system. Scientists attempt to solve problems on a flat piece of paper (two-dimensionally), seeking to establish their parameters with circumferential lines used like fences. Fences do not embrace flying birds. Systems—and ergo system parameters—are inwardly-outwardly inherently omnidimensional.

Universe is ever intensively and intertensionally pulsing and resonating, convergently-divergently, explosively-implosively, in a vast range of system frequencies, magnitudes, and chords. If we have the usual human equipment, we may be intensively tuned into, and even intertuned with, other individual, special-case human systems.

With my system law, all systems are always polyhedra, and by Euler's law, all polyhedra must consist only of corners, faces, and edges. We have here, therefore, a topologically and systemically considerate method of thinking.

Systemic thinking may be fine-tuned, like a computer program, to

FIG. 6.5 *Synergetics' Constants of the Hierarchy of Primitive, Pre-Time-Size, Omnisymmetric, Four-dimensionally Expansive and Contractive Systems.* Isotropic means "everywhere and when the same." A vector is a line of force aimed in a known angular direction in respect to an axis of reference, the length of which is the product of its mass multiplied by its velocity. The vertexes of an isotropic vector matrix are congruent with the centers of unit-radius spheres in closest packing. All of the geometrical systems below are congruently describable within the unit-length isotropic vector matrix. The isotropic vector matrix is also the unified electromagnetic and gravitational field. Its vectors are its wavelengths, and its frequencies are the number of vector-edge modules characterizing the system's topological description.

Old Name	New Name	Minisystem Volumes	Number of Polar Somethings	Number of Nonpolar Somethings		Number of Framed Views of Nothing
Tetrahedron	The "four-corner" ministructural symmetric system	1	2	2	+	4
Cubo-octahedron	Vector equilibrium twelve-corner	2½	2	10	+	20
Cube	Double tet, eight-corner	3	2	6	+	8
Octahedron	Six-corner, middle structural system	4	2	4	+	12
"Sphere"	Rhombic triacontahedron, "spheric"-icosa 242 corners, 120 "T" modules equal 5 minisystem volumes	5	2	240	+	480
Rhombic dodecahedron	Spheric domain fourteen-corner	6	2	12	+	24
Icosahedron	The "twelve-corner" maxistructural system	18.51	2	10	+	20
Cubo-octa	Vector equilibrium twelve-corner	20	2	16	+	32

Updating the earlier *Synergetics* charts as of May 1, 1983

Recognizing both the additive *twoness* of the two poles of independent spinnability of all systems and the *multiplicative twoness* of all systems,' inside concavity and outside convexity as discovered and published in *Synergetics'* topological hierarchy of primitive systems whose topology and angles are constant independent of size.

128

Number of the System's Structural (Push-Pull) Lines of Interrelationships Existing Between the System's Corner Somethings	Number of Nonpolar Somethings		Number of Views of Nothingness Framed by the System's Lines of Structural (Push-Pull) System's Corner Interrelationships		Number of Structural Interrelationships Existing Between the System Shape Defining Corner Somethings		Prime-Number Multipliers
6	1	+	2	=	3	×	1
30	1	+	2	=	3	×	5
18	1	+	2	=	3	×	3
12	1	+	2	=	3	×	2
720	1	+	2	=	3	×	5 × 3, × 24
36	1	+	2	=	3	×	3 × 2
30	1	+	2	=	3	×	5
48	1	+	2	=	3	×	2^3

The only variables are the first four prime numbers, 1, 2, 3, 5, or interproducts thereof.

[signed]

R. Buckminster Fuller

Discoverer and Copyrighter

When a cube and a square are employed, as in 1983 physics, as the universal units of volumetric and area measurements, the following incoherence and lack of mathematical integrity obtains:

	Surface area	Volume
Cube	6	1
Tetrahedron	1.7421	.1179
Octahedron	3.4641	.4714
Dodecahedron	20.6457	7.6631
Icosahedron	8.6603	2.1813

reject or correct any topological inharmonies or faulty parameters. The computer, despite the popular misconception, can answer only specifically relevant system questions. It cannot answer the question What shall I do? It can, however, answer, Of my various options, which is logically and physically most economic?

Systems powerfully and spontaneously brain-employ our inward-outward, convergent-divergent, concave-convex, size-determining, general sorting-out and concepts-differentiating capabilities. Each and every thought is a tuned-in system of uniquely interrelevant experience recalls. The images of our image-I-nation are systems and necessarily concepts as well.

Thought systems consist of all clearly relevant considerations. *Consideration* means literally bringing together and has its origins in stargazers' discovery of constellations, the interrelating of neighboring stars—*sidus* means "star," as in the word *sidereal*.

Thought systems have their spontaneously conceived macro- and microrelevant limits. There are events obviously too large and infrequent spontaneously to come under consideration, and there are events too small and/or of too high frequency of occurrence to be encompassed within our range of macro-micro parameters.

Thoughts, like television programs, have their tuned-in, always discrete, special wavelengths and frequencies. These tuned-in frequencies inherently exclude the multitude of neighboring, concurrently broadcast, but spurious signals. At the present time, irrelevant advertising commercials frequently and unfortunately do intrude upon our chosen tuned-in TV shows, but that is another matter.

WE NOTE NOW THE FACT THAT THE Greeks—with the possible exception of Democritus—mistakenly assumed that the phenomenon "solid" existed, citing the solidity of marble as an example. Through instrumentally verified experiment, we know now that the electron is relatively as remote from its nucleus as the Earth is from the Moon, given their respective diameters and spherical activity domains. We now know that there are no true solids in existence. Further, we know of nothing in Universe touching anything else.

The incorrect Greek viewpoint led Plato to offer for consideration his geometrical "solids," thinking of them as being carved from marble or wood into the shapes of cubes, octahedra, tetrahedra, icosahedra, and dodecahedra and as therefore having solid sides, which the Greeks termed *hedra*. Thus, all multifaceted objects of solid geometry became known inappropriately as polyhedra. Because we now know that no

solids exist, we must start identifying geometrical systems more logically by the number of vertexes, for which I have developed the term
polyvertexia (see Fig. 6.6).

Sir James Jeans pronounced what is to me the most sensitively
inclusive and accurate definition of science when he said, "Science is
the sincere and consistent attempt to set in order the facts of experience." Ernst Mach, the Viennese physicist whose name is celebrated in
the measurement of supersonic speed, spontaneously and specifically
elaborated on the Jeans generalization as follows: "The special case of
science known as physics is the attempt to set the facts of experience in
their most economical order."

Jeans's comprehensive science considered all types of order, such
as size or color or weight. Mach's physics had found that nature always accomplished her tasks in the most economic energy-employing
and -expending manner. His definition, which I paraphrase here, indicates much about scientific methodology: Seeking to set in most
energy-efficient (economic) order the facts of experience.

There is no identifiable experience that is less than a system. Systems
must have insideness and outsideness. Two events have only betweenness. Three events have only betweennesses. To inclusively differentiate
and identify insideness and outsideness takes a minimum of four events
to define a tune-in-able wavelength and frequency system (see Fig. 6.8).

Since I am intent upon comprehending what all experience is trying to communicate to us and since I am intent upon being consistently scientific, I have, in my sorting-out and rearranging of facts in
systemic order of relevancy, reworded for clarity Euler's topological
characteristics.

Since what I have learned is that all experiences are systems; that the
vertexes which geometrically identify systems can be, and often are,
only microtunable to nondifferentiable wavelengths and frequencies;
and that the subtunable limit condition may be heard and located but not
as yet identified as a discrete signal—what is known as static or spurious
or background radiation—I will therefore identify micro corner "somethings" as "static events" and speak of these system corner events as
"somethings," represented by the letter S. I will also henceforth reidentify the system faces (the old "hedra") as triangular window-framed
views of nothingness to be mathematically identified by the symbol Δ.

I will now identify the six most economical lines of interrelatedness
of the four static somethings as the minimally six-part set of push-pull
energy vectors structurally integrating the tetrahedron. These vectors are
the twelve (six positive, six negative) degrees of freedom coping with

Prime number	New name	Old name
2^2	four-vertexion[1]	tetrahedron
3×2	six-vertexion	octahedron
$2^2 \times 2$	eight-vertexion	2 tetrahedron (cube)
$2^2 \times 3$	twelve-vertexion	icosahedron or VE
7×2	fourteen-vertexion	rhombic dodecahedron
5×2^2	twenty-vertexion	pentagonal dodecahedron
2^5	thirty-two-vertexion	rhombic triacontahedron
31×2	sixty-two-vertexion	120 (60 + 60 −) spherical 15 great circles
$11^2 \times 2$	two-hundred-and-forty-two-vertexion	31 great-circles sphere 480 spherical right triangles

FIG. 6.6 *New identification of polyvertexia.*

[1] *Tetravertexion* (plural, *tetravertexia*) is also used in this book.

FIG. 6.7 *Underlying order in superficially seeming randomness law.* The number of interrelationships X of a given number N of "something" is

$$X = \frac{N^2 - N}{2}$$

When we look at the stars, they appear to be quite randomly scattered throughout the sky. We can say, however, that the number of direct and unique interrelationships among the stars is always given by this equation. Further, we are mathematically justified in assuming order always to be present despite the appearance of disorder. Looking at the starry skies gives us a personal sense of the order-discovering power of weightless mind and at the same time a sense of our physical body's negligible size in Universe when compared to the vast reaches of visible stars arrayed across the nighttime sky.

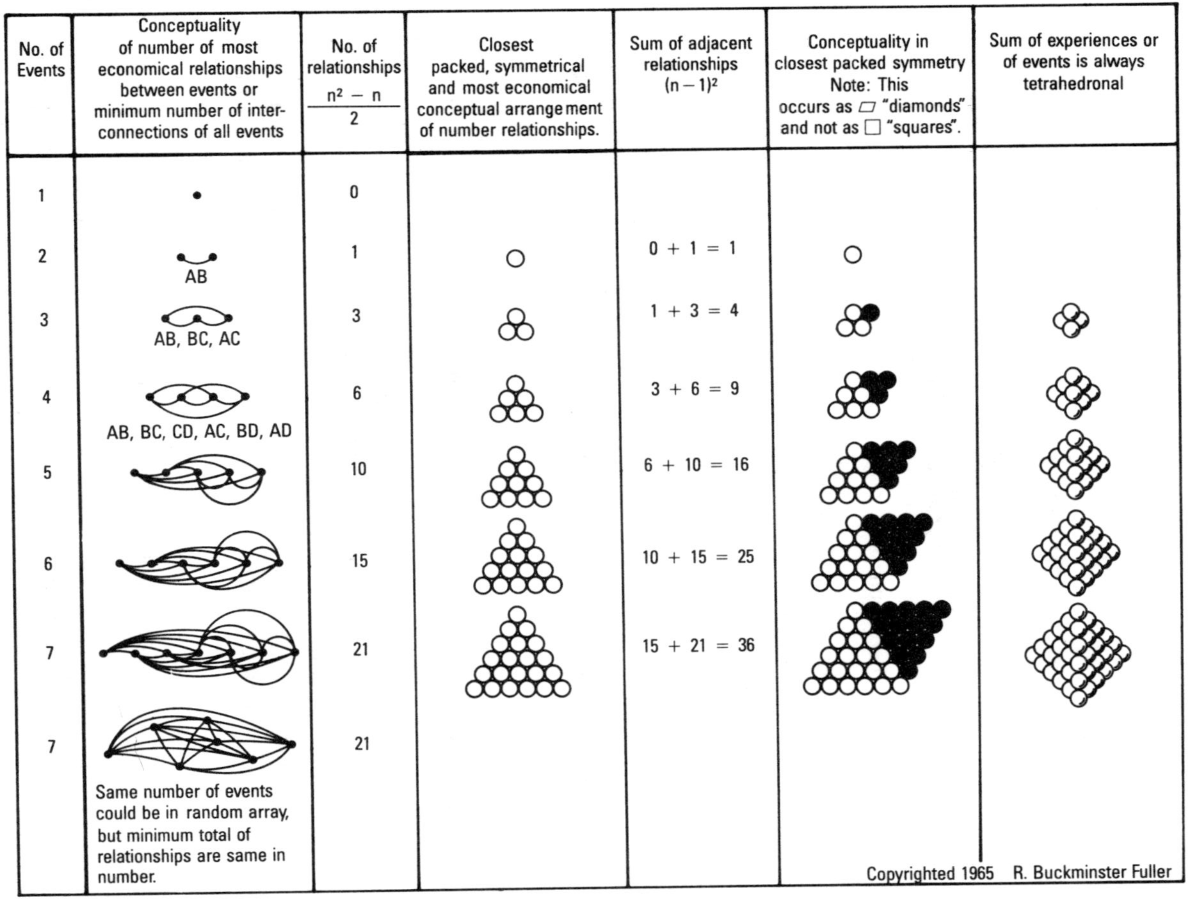

UNDERLYING ORDER IN RANDOMNESS

No. of Events	Conceptuality of number of most economical relationships between events or minimum number of inter-connections of all events	No. of relationships $\frac{n^2 - n}{2}$	Closest packed, symmetrical and most economical conceptual arrangement of number relationships.	Sum of adjacent relationships $(n-1)^2$	Conceptuality in closest packed symmetry Note: This occurs as ◇ "diamonds" and not as □ "squares".	Sum of experiences or of events is always tetrahedronal
1	•	0				
2	•⌣• AB	1	○	$0 + 1 = 1$	○	
3	AB, BC, AC	3		$1 + 3 = 4$		
4	AB, BC, CD, AC, BD, AD	6		$3 + 6 = 9$		
5		10		$6 + 10 = 16$		
6		15		$10 + 15 = 25$		
7		21		$15 + 21 = 36$		
7	Same number of events could be in random array, but minimum total of relationships are same in number.	21				

Copyrighted 1965   R. Buckminster Fuller

FIG. 6.7

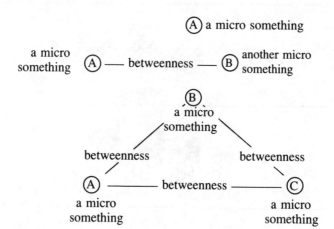

FIG. 6.8 *The minimum system.* The human-senses-tunable, differentially apprehending *minimum system* configuration of Universe has insideness and outsideness and is defined by four infra-human-senses-tunable, microsystem somethings. Each of the latter have four micro-macro something corners. Up to three relationships, as pictured above, does not constitute a system.

the structural integrity of all independently existent systems—for instance, the minimum twelve spokes necessary to stabilize the hub of a wire wheel. These twelve domains of freedom of all individual systems are those of the electromagnetic and gravitational tension and compression forces operative within each of the twelve unit-radius spheric domains that are intertangentially closest packed around any one spheric something in an aggregate of unit-radius spheres, a "sphere" being a high-frequency complex of approximately equimagnitude energy events operating at approximately equiradius distance from a center event.

Thus, scientifically corrected, Euler's equation now reads:

The number of corner events plus the number of triangular window-framed views of nothingnesses always equals the number of linear (vectorial) interrelationships of the system plus two.

This definition can, however, be improved further.

Since the most unique aspect of a system is its cosmic independence of existence derived from its twelve degrees of freedom and since all independent systems have independent rotatability, they necessarily have uniquely identifiable axes of spinnability or all-around, overall viewability and considerability.

Axes of spinnability always have two poles. We may now most economically restate Euler's topological formula of constant interrelative abundance of primitive aspects of all systems as follows:

In all polyvertexia, the two vertexially operative poles of axial spin plus the number of nonpolar vertexia plus the number of triangularly framed window views of internal nothingnesses will always equal the total number of uniquely most economical, vectorial, linear interrelationships of the system's corner vertexia "somethings."

As already noted several times, but very worth recalling, life begins with awareness. No co-occurrent otherness, no awareness. No co-occurrent otherness, no life.

One small something—too small to be described as being other than point-to-able—can be seen by another something. One something by itself, however, has no external relationships, and with no external relationships there is no life.

Note here that synergetic geometry, unlike other systems of geometry, deals with most-economical relationships (which can be called geodesics), not with shortest distances between two points—that is to say, with lines.

The only interrelatedness of two overlappingly occurrent somethings is *betweenness: AB* or *BA.*

Three simultaneously occurrent somethings have only three *betweennesses: AB, AC, BC.* (See Fig. 6.8)

Four simultaneously, overlapping occurrent somethings—*A, B, C, D*—have six betweennesses: *AB, AC, AD, BC, BD, CD.* They have an only mutually differentiated insideness and outsideness. Four somethings produce a system: a tetrahedron, the minimum differentiable something.

A *microsystem* has six degrees of freedom articulating a subtunable, subdifferentiable, complex event. A microsystem may be spoken of as a point, a blip, a static event, a spheric microsystem, or a tetrahedron so small as to make it impossible to distinguish its parts. A microsystem is an inadvertently located but not as yet discretely tuned-in static encounter.

A *minisystem* is a high-frequency, short-wavelength, discretely tuned-in, topologically identifiable system.

A point is a microsystem. A microsystem is a locatable but as yet noncomponently differentiable complex tuned in by hearing or seeing or smelling or statically touching an event.

A point *A* in our model in Fig. 6.9 is a "point-to-able" something.

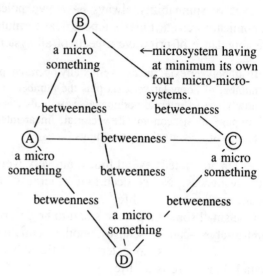

FIG. 6.9 *System outsideness*. Systems always have potentiality to be (1) discovered, (2) tuned-in microsystems inside and macrosystems outside the considered (i.e., tuned-in) system.

It is momentarily subdifferentiable, which we can also describe as the direction "in."

This static blip *A* is a something having the inherent but as yet nonsensorially differentiable insideness and outsideness of an infra-micro system enclosed by a nonidentifiable number of somethings; it is therefore not demonstrable as a simplest minimum componented system in Universe, but it nonetheless has to be a system. It has to be a subdifferentiable tetrahedron.

Unity-as-twoness is dichotomically realized in time-sequencing as the discovering of the withoutness by withinness, of the outside of self by the brain inside self, even though no humans have ever "seen" outside themselves. Humans see and realize their seeing only inside their brains (i.e., within their skulls). The information humans receive from the outside through the sense of touch has proven so consistently reliable over a period of time that the sensorial leap is made to the assumption that they are seeing the outside world, whereas in reality it is only images inside the brain that they work with. With complete accuracy, we could say to one another, "I imagine I see you sitting over there."

Inherently, there are two kinds of twoness of indivisible unit: (*a*) multiplicative twoness, (*b*) additive twoness (Fig. 6.10).

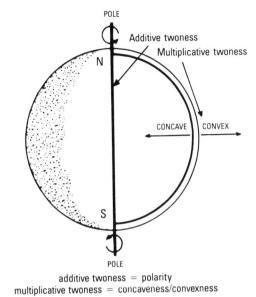

additive twoness = polarity
multiplicative twoness = concaveness/convexness

FIG. 6.10 *Additive twoness and multiplicative twoness.*

Omnidivergent or Convergent

The insideness-outsideness twoness we call the *multiplicative two-ness*. To the inside-outside twoness, the *additive twoness* is indeed "added." It is the twoness of the poles of the inherent spin axis of all inherently independent in-Universe systems. The additive twoness is the inherent polarity of our imagination's head-foot dichotomy or obverse-reverse dichotomy or of the inherent divisibility of system differentiating.

The two poles of the spin axis of observation provide all systems with time-cycling and the latter's inherent twoness of from-moment-to-moment cyclic differentiation.

Each and every thing—and ergo all things—are unique systems.

The word—the communication of an idea—is a systemic conception. The idea of greater work effectiveness through inventive-mind-elucidated cooperation made possible by speech, picture, or gesture is the initial tool of human evolution.

"In the beginning was the word," and the word was God—good, G-OO-D—i.e., two cooperative, completely individual, independent humans joined together.

Unity is plural and at minimum two. Concave and convex always and only coexist (Fig. 6.11). Concave reflectively concentrates imping-ing radiation; convex reflectively and contraction, divergent and con-

FIG. 6.11 *Yin-yang.*

vergent, and the minimum two poles of system spinnability. If unity was not inherently plural, it could not be divided to accommodate multiplication only by division into progressively larger numbers of progressively smaller systems and whole-system components. The minimum system has a minimum of twenty-eight topological components. Since multiplication is only by division, division is also accomplished only by multiplication. (See Fig. 6.12)

In electromagnetics—for instance, radio systems—there are tuned-in programs of unique wavelength and frequency, plus non-tuned in, longwave, low-frequency macroset programs of broadcast tunabilities and non-tuned-in shortwave, high-frequency microsets of broadcast programs.

Each for-the-moment *thought* has its for-the-moment relevant, tuned-in thoughts, and those tuned-in thoughts have macroirrelevant aspects that are too large and too infrequent to be considered and microirrelevant aspects that are too frequent and too short in wavelength to be conceivably relevant to the thought system considered.

All thoughts are unique systems. All thoughtful consideration and reconsideration looks for some orderly pattern to be remembered and relied upon, e.g., "Most clover has three leaves, a rare few have four leaves."

The tetrahedron, with its four corners, four faces, and six edges, is the minimum something in Universe. We have seen that we cannot break a rock into pieces that have fewer than four corners or fewer than three faces around a corner or fewer than three edges around a face.

The tetrahedron confirms Euler's formula, which, we recall, states that *the number of corners plus the number of faces of all polyhedra equals the number of edges plus the number 2.*

For a few instances:

	Corners	+	Faces	=	Edges	+	2
Cubes	8	+	6	=	12	+	2
Octahedra	6	+	8	=	12	+	2
Dodecahedra	20	+	12	=	30	+	2
Icosahedra	12	+	20	=	30	+	2

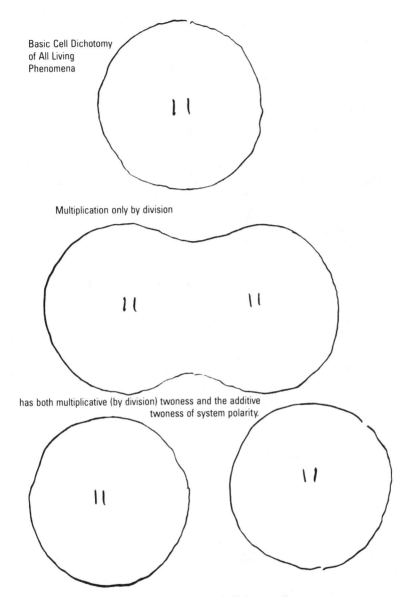

Basic Cell Dichotomy
of All Living
Phenomena

Multiplication only by division

has both multiplicative (by division) twoness and the additive
twoness of system polarity.

FIG. 6.12 *Basic dichotomy of all living phenomena.*

Edges always occur in sets of six. Edges do not exist by themselves: there cannot be an edge to nothing. Neither insideness nor outsideness exist by themselves, nor do corners.

Erstwhile "modern physics" persists in operating modellessly—and ergo blindly—with the mathematical tools of complex imaginary numbers, probability, calculus, and *XYZ*-coordinate frames of reference for plotting codiffering rates of change of experimentally evidenced statistics, in hope thereby of discovering an equation-expressible, generalizable interrelationship (a principle).

Physicists and other scientists still misassume that an *XYZ* perpendicular-parallel, three-dimensional coordinate system provides a framework of dimensional reference that can accommodate and satisfactorily express experimentally gained information interrelationships.

Experience has disclosed no solids, no straight lines, no continua, no parallels, no Greek spheres, no up and down, no absolute state of rest. Experience only discloses waves of divergent events and interference-knotted amassing of convergent events, producing only angles and frequency of angular interrelationship alterations.

All design consists entirely and solely of angle and frequency modulation. Universe operates convergently-divergently, expansively-contractively, radiantly-gravitationally, integratingly-disintegratingly, everywhere and everywhen intertransforming. Convergent-divergent Universe operates systemically, successively tuning in its overlapping scenario episodes operating between its extremes of tuned-in microcosmic-macrocosmic regional events.

Universe does not—in fact, cannot—operate as a one-dimensional, straight-line phenomenon. One-dimensionality, having neither insideness nor outsideness, cannot be conceptually embraced or experimentally evidenced. Unveering linear straightness cannot be physically demonstrated.

Nor does Universe operate as a two-dimensional, planar phenomenon having no insideness or outsideness. No such phenomenon can be experienced, conceptualized, or experimentally reproduced.

Nor does Universe operate as exclusively three-dimensional, mutually interperpendicular *XYZ,* straight-line delimited, three-way cross of parallel referencing, which, having neither insideness nor outsideness, cannot be experimentally—which is to say, experientially—demonstrated.

Demonstrable local Universe always and only operates as a convergent-divergent, nucleated, or vacantly centered insideness and outsideness system; a growable or shrinkable, spherically expandable or contractible, radiant-wave-propagatable system; a gravitational, spherically embracing, pulsatively expanding and contracting, simultaneous, four- and six-dimensional synergetic system. There are no experientially demonstrable *nonsystems,* nor are there experientially demonstrable *parts* independent of systems.

Teaching that a system can be built of parts—as is done in all schools—overlooks the fact that the parts are each systems in themselves, each dividing all Universe into everything outside the system, everything inside the system, and the system itself. We can only start experientially with system and thereafter discover the constituent parts. A system has inherently irreducible minimum aspects: its convergent aspects, which we know as vertexes; its divergent opposite-to-vertex openings, which we know as faces; and the vectors, which demonstrate most-economical energy and time interrelationships, which we know as lines (geodesics), and which also delineate and enclose. Synergetics' study of these unique aspects and their interrelationship constancies overlaps, and in many cases advances, some areas of what is known to mathematicians as topology.[2]

The three prime topological aspects can be individually emphasized while obscuring the geometrical multivertexia (formerly polyhedra) (see Fig. 6.13).

The only topological aspect clearly shown in each model is that of the vertex.

Mathematical law is eternal—exceptionlessly constant.

If I knock off one corner from any one of the regular symmetrical polyvertexia, making it irregular, the law persists.

For instance, in Fig. 6.14 one corner of the tetrahedron (four-vertexion, or tetravertexion) is knocked off, leaving in its place a small triangular facet. We have now lost one old corner (a small tetravertexial system) and have gained three new corners B', C', D' (net gain of considered system: two corners). We have also gained one additional triangular face $C'D'B'$ and three additional new edges $B'C'$, $C'D'$, and $D'B'$. The three areas $B'C'CB$, $C'D'DC$, and $B'D'DB$ are trapezoids, which are structurally unstable; to correct this, we install triangulating vectors BC', CD', DB'. After removing the small tetravertexion $AB'C'D'$, our total topological score of the remaining big truncated tetravertexion is $6V + 8F = 12E + 2$, or the total twelve structural interrelationships vectors existing between six corner somethings plus 2.

In our topological consideration, it matters not if our original tetrahedron, octahedron, or icosahedron—or thought or stone—is irregular in its angular, linear, or facial dimensions.

Euler had discovered that his topology embraced all viewable features of any system.

---

[2] Topology is qualitative (rather than quantitative) geometry that deals with order rather than size (or time).

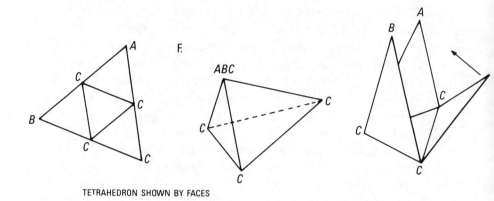

TETRAHEDRON SHOWN BY FACES

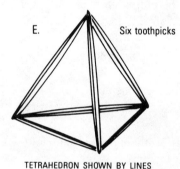

TETRAHEDRON SHOWN BY LINES

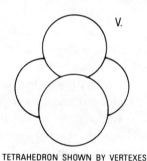

TETRAHEDRON SHOWN BY VERTEXES

FIG. 6.13 *The three ways of physically demonstrating the simplest system in Universe—the four-vertexion.* The tetrahedron (tetravertexion or four-vertexion) can be equally validly drawn as:

1. The Platonic "solid" emphasizing the four "faces," which alternatively are known in synergetics as divergent openings
2. The six "lines" or "edges," which alternatively are known in synergetics as vectors
3. The vertex domains, which alternatively are known in synergetics as closest packing of spheres

Whether it is a Rembrandt or a child's freely ranging—so-called two-dimensional—pencil drawing, you will find that the whole picture scheme always can be sorted into lines (edges), areas (faces), and crossings (vertexes, corners, or points), leaving no unaccounted features of the picture.

The points, lines, and areas may be of any color; where different colored areas abut, a line occurs. No matter how you choose to classify any feature of a Rembrandt, the formula of relative abundance of line, points, and areas will hold.

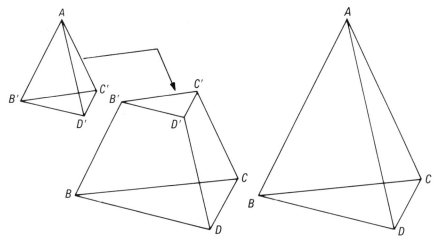

FIG. 6.14 *Tetrahedron and truncated tetrahedron.*

If you are considering only the painted-on front face of a wood-frame-mounted canvas (Fig. 6.15), you are dealing exclusively with only one face of an always-polyhedral system. Pretend as you will—and as schools encourage you to do—that you are dealing only with a two-dimensional plane, but in reality (i.e., the four-dimensional Universe), planes always and only exist as facets (faces) of polyhedral systems. Euler himself was still ensnared on academia's supposition of parts and separate dimensions having an independent existence from whole systems.

Euler played his topological game in plane geometry as with children's linear sketching, in which the number of crossings plus the number of divided-off areas always equals the number of line segments plus one. Euler himself was subject to the self-deception of an independently existent two-dimensionality reality. He, like August Möbius of Möbius-strip fame, saw the paper as having no insideness.

We know that a flat sheet of paper is always a very thin polyvertex-

FIG. 6.15 *Wood-frame-mounted canvas showing all its dimensions.*

ion with two large faces, front and back, and four extremely narrow side faces, with eight corners and twelve edges (see Fig. 6.16).

All existent and thinkaboutable otherness systems are always four-dimensional, facetwise, with the four *planes of symmetry* of the minimum system in Universe, the tetra- or four-vertexion (the old tetrahedron) and its contained hexavertexion (formerly octahedron). The hexavertexion (or six-vertexion) is also six-dimensional edgewise, as is the tetravertexion, with its six edges and the hexavertexion's twelve-edge systems (see Fig. 6.17).

If our originally broken-off (symmetrical or asymmetrical polyhedral system) rocks or stones are thrown or fall off a sea cliff, they will become progressively rolled, smitten, crushed, or nicked by local landslides or by the surf. Under such conditions their corners and edges get progressively lopped or worn off, leaving them with a progressively greater number of facets, corners, and edges. Despite irregular, asymmetrical fractionation, the constant *relative topological abundance* of corners, facets, and edges will be rigorously maintained as these independently evoluting polyhedral systems progressively get rounded off and approach a seeming smoothness.

If viewed with a microscope of adequate magnitude, rocks will always be found to be polyhedral systems. Even polishing them to superficial shininess will not prevent a microscope of sufficient magnification from revealing more and more sets of Euler's constant relative abundance of corners (points), edges (lines), and faces (areas) as given by his formula $V + F = E + 2$.

Finally, using electron microscopes, we see individual crystals and their separate, unique molecules and those molecules' separate, unique atoms. The relative interabundance of those electron, proton, and other systemic interstructurings must also always conform to Euler's relative abundance of corners, faces, and edges.

As we explore physical systems ever more inwardly (microcosmically), we observe again that the electron is as remote from its nucleus as the Earth is from the Moon, considered in respect to their relative diameters.

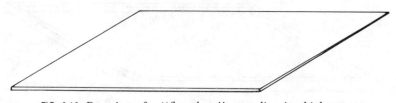

FIG. 6.16 *Drawing of a "flat plane" revealing its thickness.*

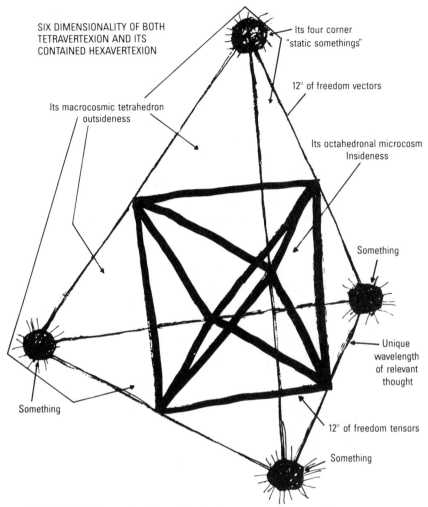

SIX DIMENSIONALITY OF BOTH
TETRAVERTEXION AND ITS
CONTAINED HEXAVERTEXION

Its four corner
"static somethings"

12° of freedom vectors

Its macrocosmic tetrahedron
outsideness

Its octahedronal microcosm
Insideness

Something

Unique
wavelength
of relevant
thought

Something

12° of freedom tensors

Something

FIG. 6.17 *Six-dimensionality of both the tetravertexion and its contained hexavertexion.* Diamonds are the minimum physical material system. Thought of tetra-octa systems are the minimum metaphysical (conceptual) system.

We go on to discover that nothing in Universe is touching anything else in either the macro- or microomniintertensioned (tuned-in) systems.

The system component intertensioning always conforms to Newton's gravitational law, which states that the relative degree of interattractiveness of any two bodies in the macro- or microcosmos always varies inversely as the second powering (n^2) of the respective arithmetical distances intervening. Halve the distance and increase the interattractiveness fourfold.

Alloying

I have introduced all the foregoing regarding primitive conceptualizing in order to elucidate the invisible microcosmic metallic alloying and the surprising increases in structural and mechanical function performances per ounce of material, erg of energy, and second of time invested in any given technological task.

We discover that the cube, which is given such structural importance by the academic and corporate world, can be proven to be nonstructural.

Twelve equilength tubes strung together with two separate and parallelly led strings, each of which emerges from a tube and is led to the ends of two different tubes, will produce a cube with eight flexible corners (Fig. 6.18). If we take the midtube points of any two parallel opposite tubes *A* and *B*, hold those tubes as far apart as the assembly will allow and parallel to the ground, and let the rest of the assembly hang from those two tubes, the assembly will take the shape known as the cube.

Gravity gives the flexible-corner cube the shape of its four square curtain walls. The assembly, however, will not stand vertically on its own structural stability. Cubical shapes in architecture require corner gussets or triangular braces to prevent the shape from sagging or distorting.

There is no inherently self-forming cubical structure occurring as a primitive polyhedron in nature. Two symmetrical tetrahedra of the same size can be interposed, however, to form a structure whose four corners can be integrated to produce a symmetrical system whose eight corners form the corners of an implied cube, but the cube's twelve edges will be lacking.

There are no solid cubes. Cubical building blocks are figments of the imagination. There do exist complex aggregates of systemic events that employ eight-corner symmetries which may be spoken of as cubical, but they are not primitive structures in their own right.

Newton's law of relative interattractiveness of any two separately paired bodies relative to the interattractiveness of any other two separately paired bodies equidistantly apart with the first pair of bodies

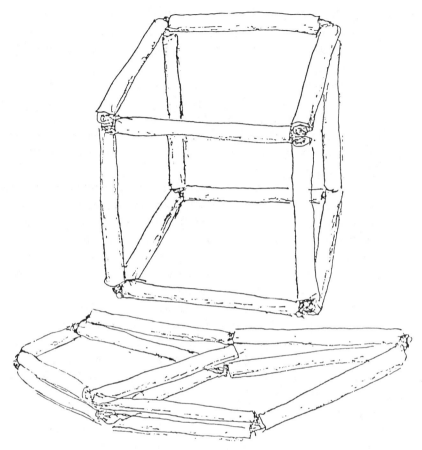

FIG. 6.18 *Flexible-corner cube.*

would be manifest as the relative magnitude of the products of the masses of each pair of bodies.

To give an example, if the first equidistant pair's individual masses are 5 and 7, and if the second equiinterdistanced pair's are 12 and 20, the respective pair's initial relative interattractivenesses would be as 35 is to 240, or 35/240.

Newton's physically, consistently proven law shows that the interattractiveness of any two bodies varies inversely as the second power of the varying arithmetical distance intervening. That is, to halve the arithmetical distance between them is to fourfold the interattractiveness. Doubling this arithmetical distance reduces the interattractiveness to one-quarter of its initial force.

In employing Newton's law to explain the tensile strengths of various nonmetallic materials, and especially the intercoherence forces of metallic alloys, we have to consider, and mathematically cope with, the convergent-divergent, four-dimensional interspacing of the system's constituent atoms.

Any of the metallic elements' symmetrical constellations of atoms may be concentrically integrated—alloyed with one or more other metallic elements' symmetrical constellations of atoms—only when they all together combine in a configuration of greater complexity which is overall an omnisymmetrical, gravitationally or electromagnetically interattractively cohered constellation.

The simplest of omnisymmetrical elemental constellations is that of the regular tetravertexion—formerly known as the tetrahedron. Assuming the individual atom to be conceptually illustratable as a superficially spherical, resonantly purring, pulsating, occulting complex of great-circle whirring events operative in pure principle, Fig. 6.19 illustrates what we mean by the minimum omnisymmetrical constellation—the tetrastellar or tetravertexial constellation.

To illustrate alloying, I employ two tetravertexia, the simplest of all symmetrical atomic constellations. I designate these two tetravertexia "red" and "blue." To produce the red tetravertexion, we take four balls of equal radius A, B, C, and D, each representing an atom (a complexedly interbalanced, microconvergent energy locus). The six edges of this tetravertexion represent vector-tensors of equal length. Because each of the six edges is a push-pull vector (or energy-force magnitude) of equal length, the forces balance and together produce the structural integrity of the system. The blue tetravertexion is designated W, X, Y, and Z (see Fig. 6.19).

These two four-ball tetravertexion systems can now be brought together in such a symmetrical manner that their centers of volume are congruent and the centers of their eight balls will coincide with the eight corners of what was formerly thought of as a regular cube.

We take the midpoints of each edge of the red and the blue tetravertexia and interpose the two tetravertexia in such a way that the midpoints of each tetravertexion are congruent with the six midpoints of the other (see Fig. 6.23). (This midpoints may be interconnected to form an octahedron, which we call a sexvertexion.)

Looking at one square face AWDX of the cube in Fig. 6.23, we have a condition where the original distance between any two corner balls of red tetravertexion ABCD would all be the same as AD, and the original distance between any two corner balls of blue tetravertexion XWYZ

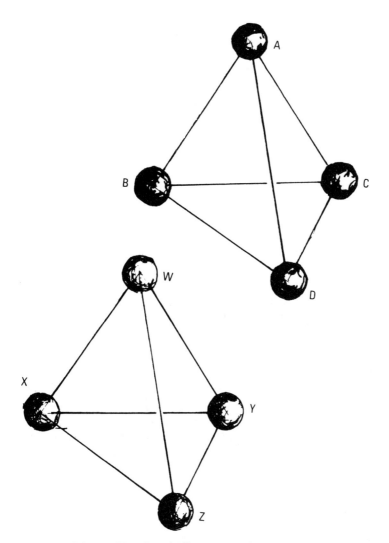

FIG. 6.19 *Two four-ball tetravertexion systems.*

would all be identical not only with one another but with the distances between any two of the four diagonally opposite corner balls of the positive tetravertexion *ABCD.* In the square *AWDX,* the uniform inter-distancing of either of the two tetravertexion's red balls or blue balls is seen to be that of either of the diagonals *AD* or *XW.*

Now, however, we note that in Fig. 6.23 *A*'s nearest neighboring ball is no longer *D* but instead *X* or *W* or *Y. AD* is the hypotenuse of the

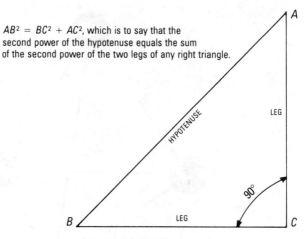

$AB^2 = BC^2 + AC^2$, which is to say that the second power of the hypotenuse equals the sum of the second power of the two legs of any right triangle.

FIG. 6.20 *The right triangle.*

right-angled triangle *AWD*, and *AW* and *DW* are the equilengthed legs of the isosceles right triangle *AWD*. Recalling the oft-proven geometrical proposition that the sum of the second powers of the two sides of a right triangle equal the second power of the hypotenuse (see Fig. 6.20), we assume the distance AD $=\sqrt{2}=$ 1.414214, and then the distances *AW* or *DW* each equal 1, wherefore *A* and *D* in their tetravertexion relationship are 1.414214 apart from one another. In this cubical arrangement, *A*'s nearest neighbors are only a distance of 1 away.

In respect to our two separate red and blue tetravertexia *ABCD* and *WXYZ*, let us assume that each of their corner-ball masses equals 1, the relative integral interattractiveness magnitude of any two of the *ABCD*'s red balls or of the *WXYZ*'s blue balls would also be exactly the $\sqrt{2}$, which is 1.414214.

When we push the red and blue tetrahedra together in the manner previously described, we now find that the distance between the complex eight-corner-ball system's nearest neighbors has been reduced from 1.414214 to 1. (See Figs. 6.21 and 6.22)

Now we show below the general mathematical expression of Newton's law and the substitution in it of the special case of our "star cube" of paired red and blue identical tetravertexial constellations of four equimass vertexial balls.

With reference to Fig. 6.23, our special case can be reduced to the following statement: Force between *A* and *D* we will call *f*; force between *X* and *D* we will call *f'*. Thus:

$$: \frac{(\text{constant})\ (m^1 m^2)}{d^2} \qquad f' = \frac{(\text{constant})\ (m^3 m^4)}{(d')^2}$$

$$: \frac{(\text{constant})\ (m^1 m^2)}{(1.414\ d')^2} \qquad f' = \frac{(\text{constant})\ (m^3 m^4)}{(d')^2}$$

$$: \frac{(\text{constant})\ (m^1 m^2)}{(d')^2} \qquad f' = \frac{(\text{constant})\ (m^3 m^4)}{(d')^2}$$

$$f = 2 \qquad f = \tfrac{1}{2}\,f'$$

$$: \frac{f'}{f} \qquad f' = \frac{\dfrac{(\text{constant})\ (m^3 m^4)}{2\ (d')^2}}{\dfrac{(\text{constant})\ (m^1 m^2)}{2\ (d')^2}}$$

if masses equal

$$: \frac{f'}{f} \qquad f' = \frac{\dfrac{1}{(d')^2}}{\dfrac{1}{2\ (d')^2}} = 2$$

A simple version follows:

d = diagonal = DA

d' = edge = XD

d = 1.414 d' (because of geometry of isosceles right triangle)

f = force between D and A if masses are constant

f' = force between X and D

$$f = \frac{\text{constant}}{d^2} \qquad f' = \frac{\text{constant}}{(d')^2}$$

then the ratio of force diagonal to force of edge

$$\frac{f}{f'} = \frac{\dfrac{\text{constant}}{d^2}}{\dfrac{\text{constant}}{(d')^2}} = \frac{(d')^2}{d^2} = \frac{(d')^2}{(1.414\ d')^2} = \frac{1}{(1.414)^2} = \frac{1}{2}$$

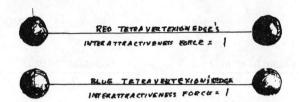

RED TETRAVERTEXION EDGE's
INTERATTRACTIVENESS FORCE = 1

BLUE TETRAVERTEXION EDGE
INTERATTRACTIVENESS FORCE = 1

TETRAVERTEXION OF
FREQUENCY 2
INTERATTRACTIVENESS
FORCE = 4

BY NEWTON'S GRAVITY ACCOUNTING LAW
THE INTERATTRACTIVENESS OF TWO REMOTE BODIES
VARIES INVERSELY AS THE SECOND POWER
OF THE ARITHMETICAL DISTANCES INTERVENING
GIVEN

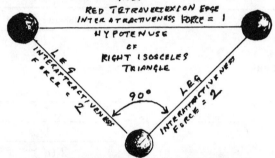

RED TETRAVERTEXION EDGE
INTERATTRACTIVENESS FORCE = 1
HYPOTENUSE
OF
RIGHT ISOSCELES
TRIANGLE

90°

LEG INTERATTRACTIVENESS FORCE = 2
LEG INTERATTRACTIVENESS FORCE = 2

BY THE SNYDER-FULLER LAW: IN A GIVEN SIZE
RIGHT ISOSCELES TRIANGLE FORMED BY THREE
REMOTE BODIES THE RELATIVE INTERATTRACTIVENESS
F THOSE TWO ONE LEG OF THE TRIANGLE APART
IS TWICE THAT OF THE SAME TRIANGLE'S
TWO BODIES ONE HYPOTENUSE APART

THE FORMAL MATHEMATICAL
CALCULATION IN PROOF OF THE
SNYDER-FULLER LAW SHOWN ON FOLLOWING
PAGES.

FIG. 6.21 *Snyder-Fuller[3] interattraction law.*

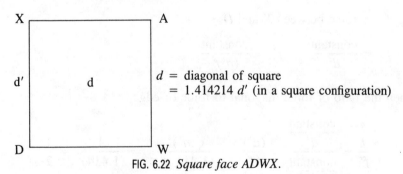

X _____ A

d'   d

D _____ W

d = diagonal of square
= 1.414214 d' (in a square configuration)

FIG. 6.22 *Square face ADWX.*

[3] Jaime Snyder [Fuller's grandson], a student of physics, consulted on the formulation of this law.

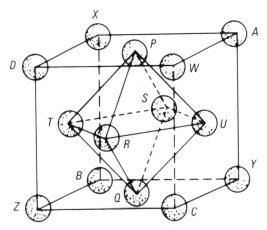

FIG. 6.23 *Intraposed tetrahedra ABCDWXYZ. Internal octahedron PQRSTU.*

From the foregoing, it is learned that in the special case of an isosceles right triangle with three equal-mass balls centered at each of the triangle's three vertexes, the interattractiveness of the pair of balls (one leg of the right triangle apart) is twice that of the pair of balls (one hypotenuse of the right triangle apart).

In our special-case consideration of triangle *AWD* of square face *AWDX* of cubical intermarriage of red tetravertexion *ABCD* with blue tetravertexion *WXYZ*, we find that the intermarriage produces a doubling of the interattractiveness between the eight balls' nearest cube-edge neighbors while still maintaining all their original greater-distance tetra-edge (hypotenuse) interattractiveness.

We may now consider an additional interallowable aspect of our red and blue tetravertexion systems: by interconnecting their mid-vector-edge crossing points, which interconnection lines describe the six-vertexion (octahedron) *PQRSTU* (Fig. 6.23).

The six-vertexion (octahedra) *PQRSTU* has six vertexes *P, Q, R, S, T, U.* The six-vertexion system *PQRSTU* has eight triangular openings or windows, which we alternately color red and blue (Figs. 6.24 and 6.25). This yields four red windows *PTR, RUQ, STQ, PSU,* and four blue windows *PUR, SUQ, TRQ, PST.* The six-vertexion *PQRSTU* has twelve vector edges *PR, PS, RT, RU, QR, QS, QT, QU, TP, TS, SU, UP.*

We may now assume that we have another six-vertexed atomic constellation *PQRSTU,* whose six vertexially centered balls are of equi-mass with those balls of the red and blue tetravertex constella-

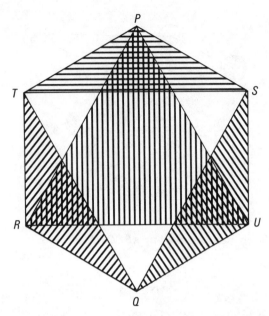

FIG. 6.24 *Alternating red and blue windows*. Red alternates in this illustration are left open for simplification of conceptualization.

tions, and that six-vertex octahedron *PQRSTU* is concentric with the star cube.

The square face *XAWD* (Fig. 6.26) will now have ball *P* at its center; ergo, balls *X, A, W,* and *D*'s nearest neighbor will now be *P* of face *XAWD* and ball *U* of cube face *WAYC* and ball *R* of cube face *DWZC* and ball *S* of cube face *XAYB*. Each of these new nearest neighbors is one leg of the isosceles right triangle *APX* away from them, whereas their former nearest neighbors had been the right triangle *APX*'s hypotenuse *AX* apart, wherefore their newer neighbors attract them twice as powerfully as had their previous neighbors, which previous neighbors had been interattracting themselves twice as powerfully as had their original neighbors. All of this double-doubling of interattractiveness did not cancel out the previous interattractiveness forces of the more remote sets of balls.

We can now appreciate how swiftly the interalloying symmetry of various atomic constellations intermultiplies their overall coherence.

In this manner alone can we understand that metallurgical alloying is not at all like the melting-together of components to make candy. In this manner alone can we understand the invisible and unexpected behavior of more performance with fewer pounds of material, ergs of energy, and

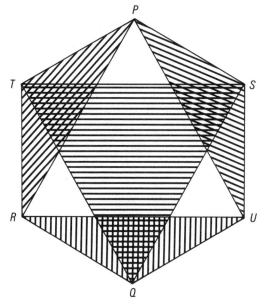

FIG. 6.25 *The blue alternates.*

seconds of time invested that now altogether have altered humanity's survival circumstances.

Only in this way can we come to comprehend why chrome-nickel-steel, whose components' tensile strengths, respectively, are 60,000, 70,000, and 80,000 pounds per square inch, produce an alloyed-together tensile strength of 350,000 psi, which is 140,000 psi greater tensile strength than the sum of those component tensile strengths, which is only 210,000 psi.

We will now mount the red triwindow *PUS* of the red tetravertex system *APUS* on the six-vertexion's red triwindow *PUS,* and the red

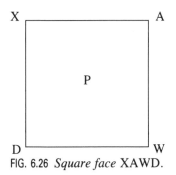

FIG. 6.26 *Square face* XAWD.

triwindow *QRU* of the six-vertexion, and red triwindow *QTS*, and finally the red triwindow *TRP* of the six-vertexion, and we will now have the "star cube" marriage of the large red four-vertexion *ABCD* with the large blue four-vertexion *WXYZ* and both concentric with the six-vertexion *PQRSTU*.

Because all the interrelationship vectorial edge lines of both the large and small four-vertexia and the six-vertexion are all constructed of equal lengths, the eight vertices *A, B, C, D, W, X, Y, Z* are all equidistant from one another and are omnisymmetrically interarrayed with all their angles equal, and the eight points *A, B, C, D, W, X, Y, Z* describe the corners of a quasicube (Fig. 6.27). We say quasicube because there is no vectorially triangulated stable cubical structure. The cube is a superficial shape resultant upon a complex of a priori structural events.

With the foregoing alloying interaugmentation of omnisymmetrical vectorial omniintertriangulated constellar system structuring, we can

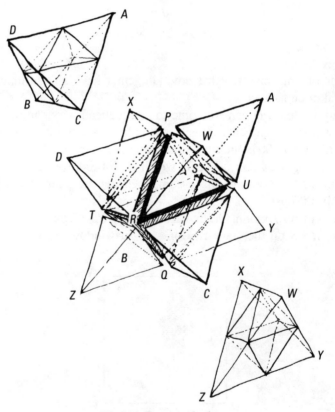

FIG. 6.27 *Star octahedron.*

well appreciate the multifold increase in system cohesiveness that is occasioned by the introduction of only one more atomic sphere M at the center of our quasicubical, comprehensive, alloyed system, as can be seen in Fig. 6.28.

Cubes have long been thought of as allspace fillers, because the Greeks found that a large cube could be subdivided into smaller cubes to reconstitute the original cube. It also seemed roughly provable that if similar-sized cubes were stacked on a true plane surface, they would fill all cubical space. But, having proven centuries ago that we live on the surface of a sphere, how is a true plane surface to be achieved?

It has been found in everyday practice that when rectilinear boxes are stacked vertically, the upper boxes have an irrepressible tendency to lean apart or fall away from one another. This has been explained, and

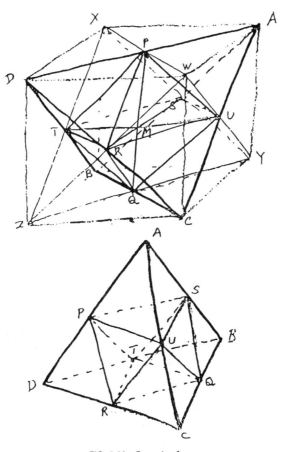

FIG. 6.28 *Quasicube.*

wrongly so, as being caused by the friction and inertia of the bottom boxes, the cumulative weight of the pile, the springiness of the box materials, and sundry other spurious reasons.

The real reason the tops of stacks of vertically stacked cubes come apart is because the Earth on which we live and vertically stack our cubes is a "spheric" system surface, and no two perpendiculars to a sphere are ever parallel to one another. Stacked vertically outwardly from the Earth's surface, the cubes are inherently, if minutely, radially divergent. Suspended inwardly in a well, they are radially convergent (see Fig. 6.29).

When builders' bubble-centered spirit levels are used to produce cement floors, those floor surfaces, as with large, smooth ice ponds, become inherently local segments of the planet Earth's spherical sur-

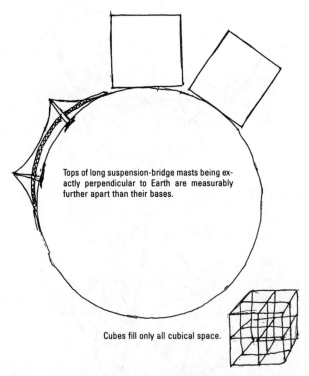

Tops of long suspension-bridge masts being exactly perpendicular to Earth are measurably further apart than their bases.

Cubes fill only all cubical space.

FIG. 6.29 *Earth with apparent perpendiculars on surface shown to diverge.* Tops of long suspension-bridge masts, being exactly perpendicular to Earth, are measurably farther apart from each other than are their bases. Cubes fill only all cubical space.

face. That is why the tops of floor-stacked vertical columns of rectilinear containers tend to rock apart. (See Fig. 6.31)

Twelve Around One

Because there are no solids in Universe, there cannot exist any solid spheres—which solidity the Greek definition of a sphere necessitated. We now know that the seeming spheric experience is always that of experiencing a polyvertexion of very high frequency. Further use of the word *sphere* in this discourse will always refer to a high-frequency polyvertexion.

Twelve spheres of uniform radius can be closest packed around 1

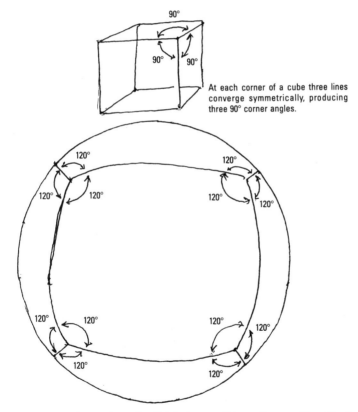

FIG. 6.30 *The spherical "cube."* It is impossible to "square" or "cube" a sphere. Since we live on a sphere in an omnicurvilinear operative Universe, it is futile to mensurate squarely and cubically. All we do are "squares."

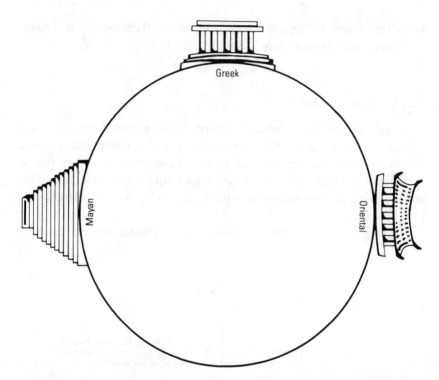

FIG. 6.31 *Earth surface considerations around the world.* Greek temple builders used plumb bobs, and their temple steps, if longitudinally sighted, will be found to be inadvertently following the curvature of the Earth. Mayan foundations were correctly engineered to be tangent to Earth and were conscious of the planet's spherical surface curvature. Many buildings in Asia were derived from ships drawn up on land; thus, their lines are reflection patterns of a ship's lines.

sphere. Spheres can be closest packed around 1 sphere in layer after layer outward ad infinitum. Each layer will always consist of six square and eight triangular facetings. The first layer has 12 spheres; the second layer, 42; the third layer, 92; the fourth layer, 163; and the fifth layer, 252. The number of each successive outwardly closest-packed surroundment will always be modular frequency to the second power multiplied by 10 plus the number 2, which is written as $10F^2 + 2$.

A spheric is not a sphere. A spheric is a high-frequency polyhedron whose corners are at approximately the same radius from the polyhedron's center (Fig. 6.32). Thus:

1. A single spheric microsystem (a six-degrees-of-freedom event complex microsystem) is free to rotate in any direction.

2. Two tangent spherics are free to rotate in any direction, but must do so cooperatively. They are friction-geared together.

3. Three omniintertangent spherics can rotate cooperatively only about their three intertangent axes, which are parallel to the edges of the equiangled triangle defined by joining the sphere centers. Thus, if the top of each spheric rotates inwardly toward the center of the triangle, then the bottoms of all three spherics rotate outwardly. This produces a top involuting and bottom evoluting pattern.

4. Four inter-closest-packed spherics block any turns or other motion of any of the four, and their interstabilized pattern produces a *structurally stable* system. Taken together the four spherics have insideness and outsideness. Each corner spheric is a complex microsystem. The four together constitute a minimum system. No rotation is possible, making it the minimum stable closest-packed spheric system: the tetrahedron.

5. The four spherics can be of different radii and will interarrest one another's motions, provided the smallest sphere's radius is such that it is too large to permit it to roll through the opening between the three largest spherics.

6. All systems have their unique wavelengths of the radii of the system.

7. Every system has an inherent (*a*) center of volume, (*b*) axis of spin, and (*c*) average radius, at whose center of volume occurs the turnaround from convergence to divergence, from contraction to expansion, from implosion to explosion, from incasting to outcasting, from tuning in to tuning out.

8. A vector is a line representing an operative energy. Its length equals the product of the mass and velocity involved in a given direction.

9. Every system has six positive and six negative vectors. These twelve, half of them positive and half of them negative, can be paired into six interstabilized, push-pull, structural components.

10. The push-pull, paired vector structural system shapers are also the system "edges" or "lines" of the mathematician Euler's three basic conceptual, topological components in his "polyhedral" formula $V + F = E + 2$. The paired push-pull vectors are the E's.

It is this principle of omniembracing, omnidirectional, twelve-unit-radius spheric systems around one spheric system that governs all convergent-divergent experience and thinking and accounts for the inherent twelve degrees of freedom that must be coped with in all independent-system internal structuring and the separating-out of an in-

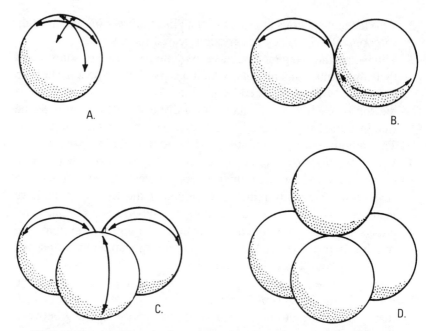

FIG. 6.32 *Four spheres lock as a tetrahedron.* Four unit-radius spheric "somethings" (microsystems) when closest interpacked form a tetrahedron.

A. A single spheric microsystem (a six-degrees-of-freedom event complex microsystem) is free to rotate in any direction.
B. Two tangent spherics are free to rotate in any direction, but must do so cooperatively. They are friction-geared together.
C. Three omniintertangent spherics can rotate cooperatively only about their three intertangent axes, which are parallel to the edges of the equiangled triangle defined by joining the sphere centers. Thus, if the top of each spheric rotates inwardly toward the center of the triangle, then the bottoms of all three spherics rotate outwardly. This produces a top involuting and bottom evoluting pattern.
D. Four inter-closest-packed spherics block any turns or other motion of any of the four, and their interstabilized pattern produces a structurally stable system. Altogether, the four spherics have insideness and outsideness. Each corner spheric is a complex microsystem. The four together constitute a minimum system. No rotation is possible, making it the minimum stable closest-packed spheric system: the tetrahedron.

dividual system within a more complex system. (See Figs. 6.33–6.36.)

"Spheric experiences" can be of three kinds: (1) polyvertexia single-bounded vertex to vertex as gases occupying maximum space; (2) double-bonded as liquids edge to edge, occupying less space than the single-bonded gases; and (3) triple-bonded as crystals occupying the least space. Since nothing in Universe touches anything else and is remotely cohered as single-bonded gases are only gravitationally, tensegrity intercohered.

[*Adjuvant's note: The following passage, written six weeks before his death, is Fuller's last known writing, and as such, and also because of its revelatory nature, it is quoted in its entirety.*]

The discovery today, Sunday, May 15, at the Good Samaritan Hospital in Los Angeles [while attending to his wife], between 3 P.M. and 4 P.M., of the necessity to think realistically and structurally only in terms of the nonexistence of spheres and therefore to think only in terms of polyvertexia. This brought about the necessity of realizing that "closest-packed unit-radius spheres" of the isotropic vector matrix are always polyvertexia in different orientations with their system centers congruent with the isotropic vector matrices' vertexes but with their external structures not touching each other. These different system states (Willard Gibbs's gases, liquids, and crystallines) had different orientations, ergo three different system radii, i.e., (*a*) when situate closest to one another but not touching vertex-to-vertex, they are single-bonded as gases; (*b*) anywhen next most remotely intersituate they are edge-to-edge double-bonded as liquids; and (*c*) most remotely and as yet evenly intersituated they are face-to-face, i.e., triple-bonded as the crystalline phase of physical state (see Fig. 6.37).

Ergo, since two polyvertexia's vertexial events cannot occupy the same space at the same time, the two outermost vertexes of each of the two single-vertex-interbonding polyvertexia are not congruent but are at critical proximity distance from one another to accommodate their respective gaseous system integrity states. The single-bonded gaseous phase of "spherics" are not congruent and must be spaces apart, and are only intercohered by Newton's law [see tensegrity discussion in section on Fuller-Snyder law, Fig. 6.21].

This brings us to Boyle's [Avogadro's] law: "Under identical conditions of heat and pressure, the same number of molecules of all gases of all elements will always occupy the same volume." But Boyle's [Avogadro's] law does not say how closely to one another the molecules must be situate within the given volume.

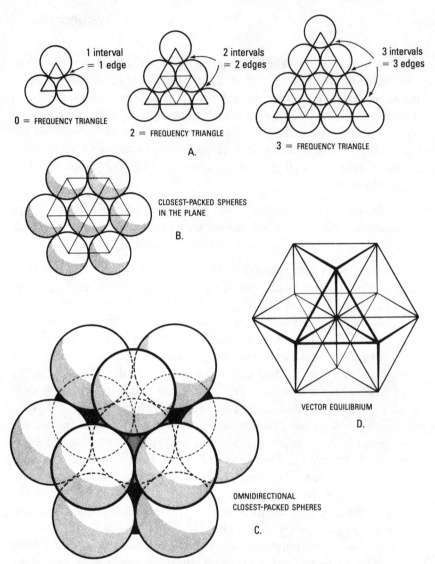

1 interval = 1 edge

0 = FREQUENCY TRIANGLE

2 intervals = 2 edges

2 = FREQUENCY TRIANGLE

A.

3 intervals = 3 edges

3 = FREQUENCY TRIANGLE

CLOSEST-PACKED SPHERES
IN THE PLANE

B.

VECTOR EQUILIBRIUM

D.

OMNIDIRECTIONAL
CLOSEST-PACKED SPHERES

C.

FIG. 6.33 *Vector equilibrium: omnidirectional closest packing around a nucleus.* Triangles can be subdivided into greater and greater numbers of similar units. The number of modular subdivisions along any edge can be referred to as the frequency of a given triangle. In triangular grids each vertex may be expanded to become a circle or sphere showing the inherent relationship between closest-packed spheres and triangulation. The frequency of triangular arrays of spheres in the plane is determined by counting the number of intervals (*A*) rather than the number of spheres on a given edge. In the case of concentric packages or spheres around a nucleus the frequency of a given system can either be the edge subdivision or the number of concentric shells or layers. Concentric packings in the plane give rise to hexagonal arrays (*B*), and omnidirectional closest packing or an equal sphere around a nucleus (*C*) gives rise to the vector equilibrium (*D*).

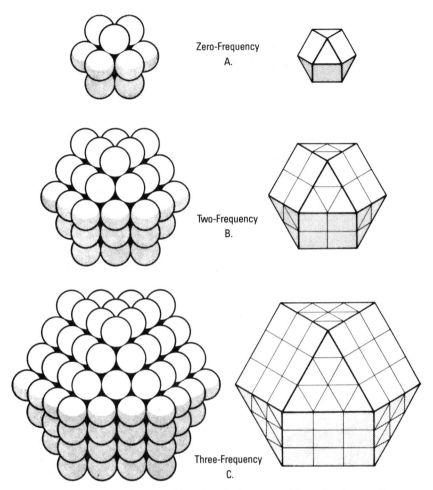

FIG. 6.34 *Equation for omnidirectional closest packing of spheres.* Omnidirectional concentric closest packings of equal spheres about a nuclear sphere form series of vector equilibria of progressively higher frequencies. The number of spheres or vertexes on any symmetrically concentric shell or layer is given by the equation $10F^2 + 2$, where F = frequency. The frequency can be considered as the number of layers (concentric shells or radius) or the number of edge modules on the vector equilibrium. A 1-frequency sphere-packing system has 12 spheres on the outer layer (*A*) and a 1-frequency vector equilibrium has 12 vertexes. If another layer of spheres is packed around the 1-frequency system, exactly 42 additional spheres are required to make this a 2-frequency system (*B*). If still another layer of spheres is added to the 2-frequency system, exactly 92 additional spheres are required to make the 3-frequency system (*C*). A 4-frequency system will have 162 spheres on its outer layer. A 5-frequency system will have 252 spheres on its outer layer, etc.

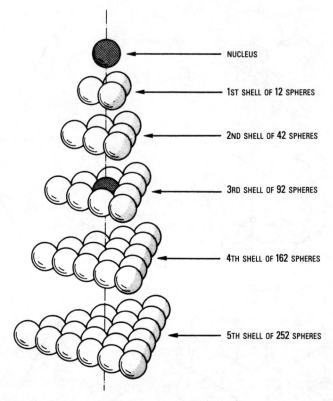

NUCLEUS

1ST SHELL OF 12 SPHERES

2ND SHELL OF 42 SPHERES

3RD SHELL OF 92 SPHERES

4TH SHELL OF 162 SPHERES

5TH SHELL OF 252 SPHERES

FIG. 6.35 *Realized nucleus appears at fifth shell layer.* In concentric closest packing of successive shell layers, potential nuclei appear at the third shell layer, but they are not realized until surrounded by two shells at the fifth layer.

This brings us also to Willard Gibbs's phase rule governing the number of degrees of freedom or energy behavior permissions necessary for its glass of ice water's water vapor, its water, and its ice to come together as the same phase and thus to occupy the same volume or space in Universe.

Gibbs's phase rule reminds us that the present-day physicist's unit of volumetric measure is that of the cube of water one centimeter to the edge at a given temperature (due to the expansion and contraction between gaseous, liquid, and crystalline phases of matter).

All foregoing discoveries, thoughts, and accounting lead to the intuitive holding on to the volumetric relationship of the spherical "fiveness" relative to the rhombic dodecahedron's sixness within which our yesterday's "unit-radius spheres" were misconceptioned to be tangen-

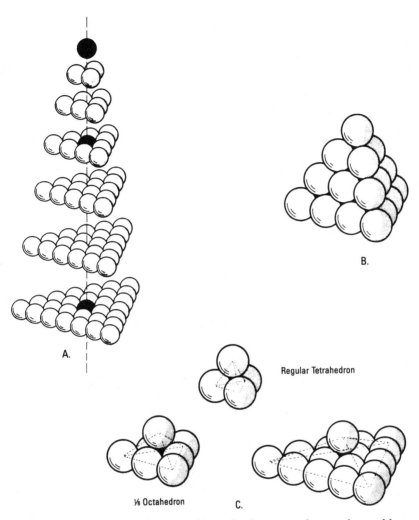

FIG. 6.36 *Tetrahedral closest packing of spheres: nucleus and nestable configurations.*

A. In any number of successive planar layers of tetrahedrally organized sphere packings, every third triangular layer has a sphere, at its centroid (a nucleus). The 36-sphere tetrahedron with 5 spheres on an edge (four-frequency tetrahedron) is the lowest-frequency tetrahedron system with a central nuclear sphere.

B. The three-frequency tetrahedron is the highest frequency without a nucleus sphere.

C. Basic "nestable" possibilities show how the regular tetrahedron, the ¼-tetrahedron and the ⅛-octahedron may be defined with sets of closest-packed spheres. Note that this "nesting" is only possible on triangular arrays which have no sphere at their respective centroids.

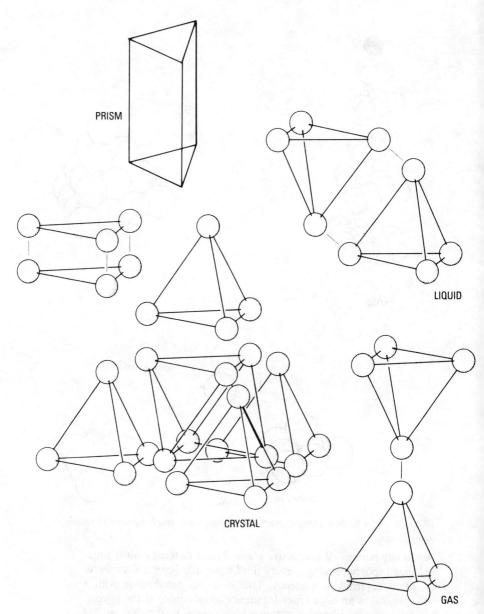

PRISM

LIQUID

CRYSTAL

GAS

FIG. 6.37 *Trivalent bonding of vertexial spheres forms rigid structures.* At *C* gases are monovalent, single-bonded, omniflexible, with inadequate interattraction, separatist, compressible. At *B* liquids are bivalent, double-bonded, hinged, flexible, with viscous integrity. At *A* rigids are trivalent, triple-bonded, rigid, with highest tension coherence.

tially situated and which "spheres" were wrongly thought of only as solids.

We now realize that the polyvertexia are single-bonded as gases, and in fact are remote from one another, and only tensegrity intercoherence has greater possible radius and lesser radiuses when double-bonded as liquid and is of lesser radius again when crystallinely phased (which explains why Planck's constant is 6.625+ rather than 6.2666 . . . to correct for the cube being threefold the unit volume of the tetravertexion). And all of the foregoing make clear that all isotropic vector matrices as the framework of reference of all energy phenomena must be considered only in their greatest radius phase, i.e., its gaseous, single-bonded, vertex-to-vertex cohered tensegrity state. Since the sphere does not exist, 3.14159 . . . does not exist and the special-case "atom" and "molecule" spheric polyvertexion occupant of each rhombic dodeca-hedron of isotropic vector matrix referencing volume of 6 can be alter-nate "phase" and operatively reoriented within the volume 5 domain as its convergent-divergent average of its interphase "state."

[signed] Buckminster Fuller
at the 15th hour of 5/15/83, with thanks to God,
the eternal sum of all truths.

What yesterday's nonscientific mathematicians have thought of as a one-dimensional line is in fact a greatly elongated system of minuscule base. What nonscientific mathematicians have thought of as two-dimensional is in fact a very thin, large-based system. What the non-scientific mathematicians have heretofore thought of as three-dimensional, having width, breadth, and height, has no inherent insideness and outsideness; ergo, it does not separate Universe into an inside and an outside, and thus is nonsystemic and therefore nonexis-tent. The tetrahedron is the minimum conceptual or physical system.

In the language of geometry, *regular* means "omnisymmetrical." The regular tetravertexion (formerly misidentified as the tetrahedron) has fourfold symmetry: four corner vertexes opposite four equiangular windows. Therefore, the regular tetrahedron can be readily divided into four equal parts. This is done by first finding the center of volume of the regular tetravertexion. Since the volume of a tetravertexion is the prod-uct of the base times its altitude, we can take the center of volume as being one-quarter of the altitude. This one-quarter-altitude point be-comes the common apex of four one-quarter tetravertexia (see Fig. 6.40).

As we have demonstrated, in contradistinction to cubes, unit-radius

The modular "frequency" of "spheric," omnidirectionally, omni-closest-packed uniform radius spheres is determined by the number of spaces between the spheres along one edge of the closest-packed system.

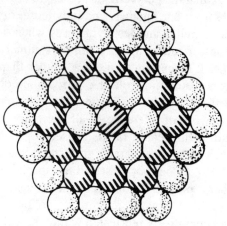

This is a three-frequency four-dimensional system of closest-packed-together unit-radius spheres. Pictured here is an equatorial layer through the aggregate at the nuclear sphere level.

FIG. 6.38 *Frequency pictured as equatorial layer through nuclear sphere.* The modular frequency of the spheric, omnidirectionally, omni-closest-packed uniform-radius spheres is determined by the number of spaces between the spheres along one edge of the closest-packed system. This is a three-frequency, four-dimensional system of closest-packed-together unit-radius spheres, pictured here as an equatorial layer through the aggregate at the nuclear sphere level.

spheres always close pack omniradially and omniintertangentially as twelve around each single sphere. Unit-radius spheres being closest packed together do not fill all the spaces (allspace). A uniformly shaped, complexedly concave, curvilinear space bounded by the spheric surface nestles between the only tangentially closest-packed aggregates of unit-radius spheres.

There is a symmetric, primitive geometrical system known as the rhombic dodecahedron (see Fig. 2.10). It has twelve uniformly dimensioned diamond-shaped facets. The geometrical centers of each of the rhombic dodecahedron's twelve diamond faces are exactly congruent with the twelve points of tangency of any unit-radius sphere, with its twelve uniformly radiused, closest-packed tangent neighbors in any such closest-packed aggregate of uniform-radius spheres.

Each uniform-size rhombic dodecahedron contains within it a uniform-radius sphere internally tangent to each of the twelve mid-

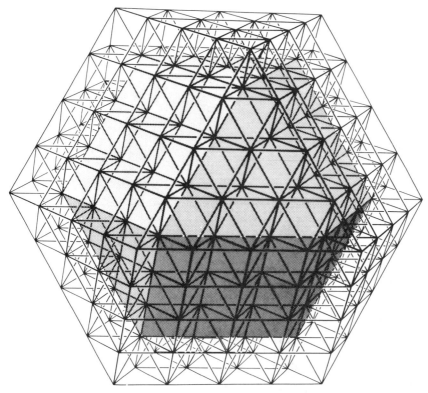

FIG. 6.39 *Nuclear structural systems.* Nuclear structural systems consist entirely of tetrahedra having a common interior vertex. They may be interiorly truncated by introducing special-case frequency, which provides chordal as well as radial modular subdivisioning of the isotropic-vector-matrix intertriangulation, while sustaining the structural rigidity of the system.

diamond faces of the rhombic dodecahedron. Uniform-size rhombic dodecahedra do closest pack, twelve around one, with one another's diamond faces exactly congruent. They interpack radially, with twelve omnidirectionally and symmetrically closest-packed around each rhombic dodecahedron in the aggregate, filling allspace. Each rhombic dodecahedron thus closest packed and filling allspace is the total domain of each of the tangentially closest-packed-together unit-radius spheres, in addition to containing the sphere.

Angle

The trails of two lines, one pre- and one post-crossing a point, or one only visibly superimposed at a distance apart from one another or a line

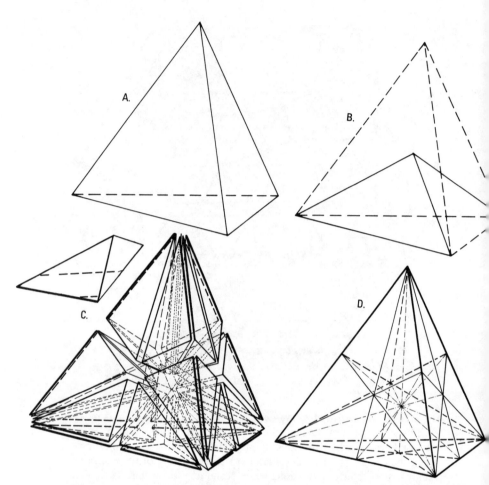

FIG. 6.40 *Tetravertexion, one-quarter tetravertexion, and one-twenty-fourth tetravertexion, or* A *module.* A, tetravertexion; B, one-quarter tetravertexion; C, one-twenty-fourth tetravertexion, which we call an A module; D, six equiangled asymmetric tetravertexia. Since one-quarter of a regular tetravertexion has been further subdivided into six similar equiangled, asymmetric tetrahedra, each of these asymmetrics is one-twenty-fourth of the regular tetravertexion. Each of these twenty-fourth subdivision tetravertexia is called an A module.

reflectively redirected or a linear wire deliberately bent, produce an angle (*V*). An angle is a visual experience—an awareness of two other-event somethings, history lines, interrelating as an angular overlay interrelationship. An angle is a conceptually imaginable interrelationship quite independent of the relative length of the angle's lines.

An angle *V* is the simplest, minimal-conceptual, attention-securing fix—ergo the mark $\sqrt{\ }$.

It takes time to measure length. Time is measured cyclically by numbers of interim completed cycles (circles). The angle is a fraction of a circle (O). Angles are subcyclic. Angles are pretime and -size conceptuality. Angles are imaginatively conceptual patterns independent of size or time (Fig. 6.41). A tetravertexion is an imaginatively conceptual structural system independent of size or time.

All systemic conceptuality that is independent of time and size we call *primitive*.

All systems and their topological characteristics are eternally true independent of size.

Tensegrity

We find that all our tune-in-able experiences are consequences of the absolute integrity of a complex family of eternal principles.

The Universe, both macro and micro, is always and only a continuously intertensioned, discontinuously compressioned structural system. It is what I call a *tensional integrity*. So often did I use that phrase that I contracted its expression to *tensegrity,* a term which has now made its way into the language. Tensegrity represents a phenomenon so universal

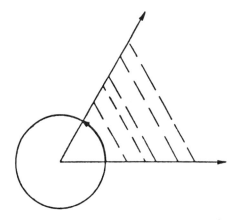

FIG. 6.41 *Angles are angles independent of the length of their edges.* Lines are "size" phenomena and unlimited in length. Angle is only a fraction of one cycle.

that it may eventually be the key to modeling a unified field theory, a tantalizing goal of the scientific community for centuries.[4]

Since nothing touches anything else in tensegrity Universe, there are no solids. What occasioned his contemporaries' conceptual acceptance of Plato's geometric "solids" was the fact that evolution had not as yet introduced humanity to exclusively tensional technology experiences and their philosophic evolutionary derivation or to the subsequently discovered electron's four- and six-dimensional gravitational integrity of interpatterning symmetries whose kinetic interstructuring behaviors produced electron microscope (nonsolid) lenses; these kinetic structuring principles in turn produced the field-emission microscope, whose lenses of abstract-principle electromagnetic integrity make possible the direct photography of one isolated atom, which single atom is in itself a complex, systemic, vector-equilibrium-referenced kinetic entity topologically omniconsistent with the eternal tensegrity principle.

When NASA was making its first rocketry experiments dealing with the problem of atmospheric reentry heat, two General Dynamics Corporation scientists were experimenting with the light, high-strength metal titanium. They made two thin-wall hemispheres of titanium sheet. One of the hemispheres had a 36-inch inside diameter and the other had a 34-inch outside diameter. They centered the 34-inch dome inside the 36-inch dome, with a 1-inch space between them, and welded a 1-inch-high titanium base ring to both the outside and inside domes. They then vacuum-pumped the air from between the two domes. Atmospheric pressure pushed the inside dome skin outward, but atmospheric pressure on the outside of the outside dome dimpled the outside dome skin inward in a pattern of hexagons and pentagons; a triangular undimpled area remained in the exact pattern of the tensegrity-geodesic icosahedron's four-frequency network. This was a *least-effort*-of-nature event and proved that nature was employing the same mathematical-geometrical logic we have been developing and considering here, showing that the icosahedron provides cosmically the *most* structurally enclosed volume per quantum of structural energy provided.

A balloon is an example of a high-frequency tensegrity sphere.

The balloon is a net with holes so small that the molecules of gas inside the balloon cannot escape. The next thing we discover is the pressure of the gases, explained by their kinetics; that is, molecules are in motion, not rigid. Nothing at all static pushes against the net. Gas

---

[4] *Tensegrity* appeared for the first time in a dictionary in 1985, in the *Oxford English Dictionary Supplement,* vol. 4, 1985, and subsequently in the *Compact Edition of the OED,* vol. 3, 1987, and other dictionaries.

molecules are hitting it like projectiles. All of the molecules of gas pressure loaded into the system are trying to get out: this is what gives the basketball its firmness. If we pump in more molecules, they become not only more crowded together but also more accelerated, producing increased heat and pressure.

The middle of the chord of an arc is always nearer to the center of the sphere than the ends of the chord. Chord ends are always pushing the net outward from the system's spherical center. Gas molecules are stretching the net outward. All outward-thrusting gas molecules have an-equal-and-opposite-thrusting-reaction molecular partner. In the tensegrity-sphere model (Fig. 6.42), each of the wooden sticks or struts represents a pair of action-reaction forces. As the gas molecules' outward caroming blows act as a total spherical enlargement network, stretching the skin, at the same time the skin (network stringing) acts to resist the outward motion (stretch). The skin is finite, closing back upon itself in all circumferential directions. All its force arrows are bound inward, balancing all the outward-bound molecules hitting the net and

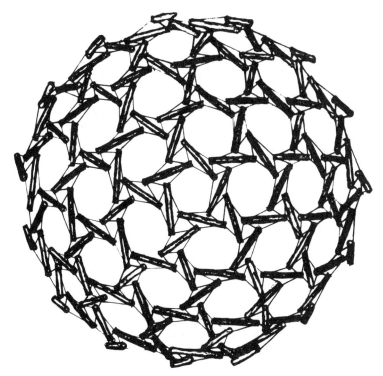

FIG. 6.42 *Six-frequency tensegrity icosahedron.*

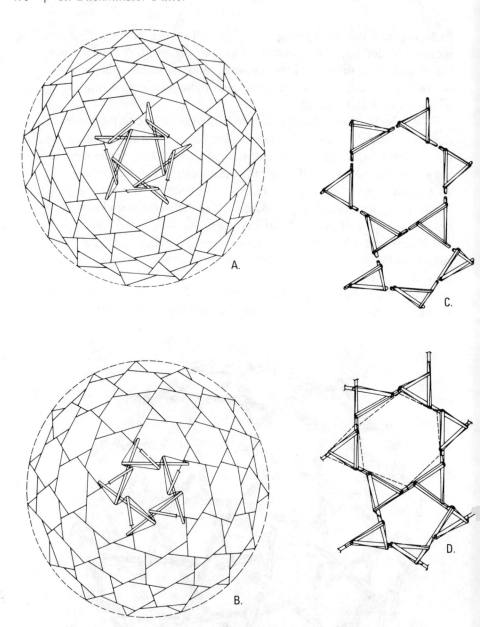

FIG. 6.43 *Single and double bonding of members in tensegrity spheres. A,*
negatively rotating triangles on a 270-strut tensegrity geodesic sphere
with double-bonded triangles; *B,* a 270-strut isotropic tensegrity geodesic sphere, with single-bonded turbo triangles forming a complex
six-frequency triacontahedron tensegrity; *C,* complex of basic three-
strut tensegrities, with axial alignment whose exterior terminals are to
be joined in single bond as 90-strut tensegrity; *D,* complex of basic
three-strut tensegrity units with exterior terminals now joined.

caroming around. Every molecular action has its equal and oppositely accelerative gas-molecule reaction mate. The paired action and reaction gas molecules produce glancing-blow, chordal-pair outward forces of the tensegrity sphere.

This is quite a different picture from that of molecules huddling together at spherical center and then simultaneously exploding outward to hit the balloon skin in an omnidirectionally outbound wave. Instead, the paired oppositely accelerated gas molecules carom around in the largest, most comfortable circles (the great circles).

All great circles cross other great circles twice in each circuit. When a third great circle crosses two others, it inherently produces six vertex crossings and eight asymmetric spherical triangles. This is the spherical octahedron. The opposite-direction-reaction molecule makes another spherical octahedron. The two spherical octahedra's twelve vertexes produce the icosahedron's twelve vertexes. Millions of these molecular events in an asymmetric icosahedral patterning average out to produce the regular icosahedral sphere.

Not only are there critical proximities that show up physically, but there are also critical proximities tensionally and critical proximities compressionally—that is, there are repellings.

What makes the net take the shape that it does is simply the molecules that happen to hit it at any one moment. Molecules that are not hitting at the moment considered have nothing to do with the balloon's or the basketball's shape. There is the certainty that other molecules might hit the network at other moments, but that is not what we are concerned with—the shape it takes at a given moment is only by virtue of the molecules that are hitting it at that moment.

Molecules near the surface of the net are coursing in chordally ricocheting great-circle patterns around the net's inner surface. Because every action has its reaction, it would be possible to pair all the molecules so that they would behave like two swimmers do who dive into a swimming tank from opposite ends, meet in the middle, and then, employing each other's inertia, bend tight their knees and bodies and shove off from each other's feet in opposite directions. This produces an acceleration effectiveness equal to what the swimmers experience when shoving off from the tank's solid wall.

This pattern indicates that if each of the paired molecules bounces off its partner and darts away in opposite directions, with each hitting the balloon net and pushing it outward with an angling blow, then to travel in a new direction but always toward the net at another point, where at critical repelling proximities each pairs off nonsimultaneously at high frequency for another repellment shove-off to ricochet off the net again, and to do so at a high event-frequency, the net will be kept stretched outwardly in all directions.

This represents what the confined gas molecules of a balloon or basketball or football or tennis ball or Ping-Pong ball are doing. With discontinuous compression and continuous tension, we make geodesic structures function in the same way.

Water always intervenes between the feet of the swimmers shoving off from one another. This water produces between the swimmers a critical proximity of their energy interpatterning.

The spaces between the energy-action-net components are smaller than are the internally captivated and mutually interrepelled gas molecules, wherefore the gas molecules, which are complex, low-frequency energy events, interfere with the higher-frequency, omnienclosing, net-webbing energy events. The pattern is similar to that of fish crowded in a spherical net and therefore running tangentially outward into the net in approximately all directions. Fish caught in nets produce an enclosure-frustrated would-be escape pattern. In tensegrities, you have gravity or electromagnetism producing the ultimate tension forces, but you do not have any strings or ultimately smallest solid threads. The more we think about it and the more we experiment, the less reliable becomes our academic concept of "solid." The balloon is indeed not only full of holes but utterly discontinuous. It is an energy network and not a bag. In fact, it is a spherical neighborhood composed of critically proximate interattractions among ultra-high-frequency energy events.

In a gas balloon, we do not have a continuous membrane of film. There is no such thing as a continuous "solid" skin or, indeed, a "solid" or a "continuous" anything in Universe. What we do have is a network pattern, a network of energy actions interspersed with vast spaces, or a lack of energy events. The mass-interattracted atomic components not only are not touching each other, but they are as relatively remote from one another as the Sun is from its planets.

People think spontaneously of a basketball as a continuous skin or a solidly impervious unitary and spherically enclosed membrane holding the gas. They say that because the gas cannot get out and because it is under pressure, the pressure makes the balloon spheroidal. This means that the gas is pushing the skin outward in all directions. People think of a solid mass of air jammed into a pneumatic bag. But if we look at this skin through a powerful microscope, we find that it is not a continuous film at all: it is full of holes. It is made up of molecules that are fairly remote from one another. It is in reality a great energy aggregate of

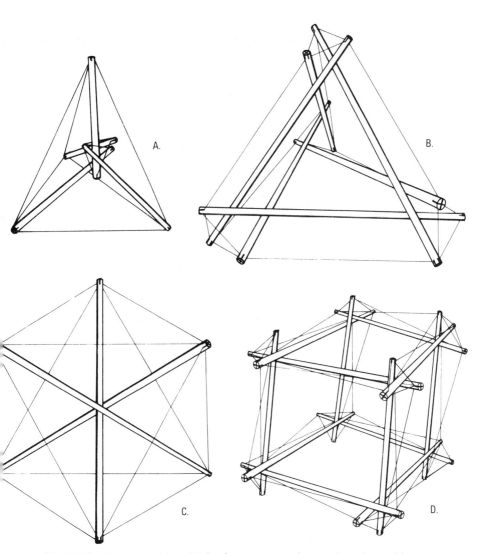

FIG. 6.44 *Basic tensegrities.* *A,* the four-strut, twelve-tendoned, outside-in (negative) tetrahedron, showing the four outer vertex turbining. *B,* the six-strut tensegrity, 18-tendoned, outside-out (positive) tetrahedron, showing central-angle turbining. *C,* the three-strut, twelve-tendoned tensegrity octahedron. The three compression struts do not touch each other as they pass at the center of the octahedron; they are held together only at their terminals by the comprehensive, triangular tension net. It is the simplest form of tensegrity. *D,* the twelve-strut, 48-tensioned tensegrity cube, which is unstable.

Milky Way–like atomic constellations, cohering only gravitationally to act as the invisible, tensional integrities of the energetic, high-frequency-event "fibers" with which the webbing of the pneumatic balloon's net is woven.

We now comprehend that geodesic tensegrity structuring provides the first true and visualizable model of pneumatic structures in which the relative thickness of the enclosing films, in proportion to diameter, rapidly decreases with the increasing size of the balloons or spheric networks.

In the case of geodesic tensegrity structures, no overcrowding of interior gas molecules, imprisoned within a submolecular mesh net, is necessary to thrust the net's structure outward from its spherical geometric center, because the compressional struts, locally islanded as outward-thrusting struts at both their ends, push the spherical net outward at every vertexial advantage of network convergence. Geodesic tensegrities are "hollowed-out" balloons that have discarded their redundantly "solid" air core. The larger the sphere, the greater the number of molecules, the lower the pressure, and the more surface on which to distribute the load or pressure impinging upon the pneumatic system. Doubling the size of the pneumatic or tensegrity sphere reduces to one-quarter the surface enclosure stress occasioned by an external force impingement of a given magnitude (see Fig. 6.45).

Geodesic tensegrities are true pneumatic structures in purest design frequency principle. They obviate the randomness and redundance characterizing the work of designers dealing only with pneumatics, who happen to be successful in blowing air into a bladder while being utterly dependent upon the subvisible behaviors of chemical phenomena. Geodesic tensegrity engineering enables discrete separation of all the structural events into two diametrically opposed magnitude classes: on the one hand, all the outward-bound phenomena, which are too large to pass through all the interstices of, on the other hand, all the inward-bound events in the too-small class. This is the same kind of redundancy that occurs in reinforced concrete, which, if drilled out wherever redundant components exist, would disclose an orderly four-prime-magnitude complex octahedron-tetrahedron truss network, disencumbered of more than 50 percent of its weight.

The geodesic tensegrity is a balloon out of which have been removed all the molecules of gas not at the moment hitting the skin and in which those specific molecules of gas that happen to be impinging from within against the skin at any one moment (thus pushing it outward) are replaced by the islanded geodesic struts; in addition, all other redundant molecules are discarded. It is possible to sew pockets on the inside

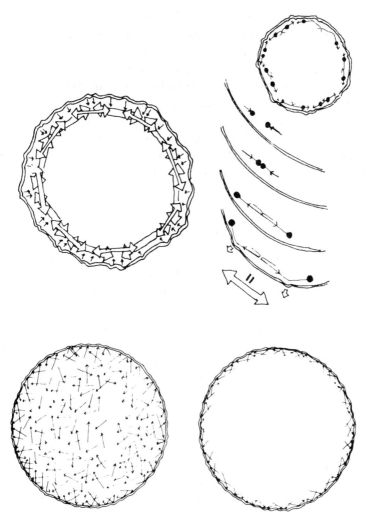

FIG. 6.45 *Chordal ricochet pattern in stretch action of a balloon net.* A gas balloon's exterior tension "net" has the shape that it has because some of the molecules are too large to escape and, crowded by the other molecules, are hitting the balloon. But the molecules do not huddle together at the center and then simultaneously explode outward to hit the balloon skin in one omnidirectionally outbound wave. The molecules near the surface are coursing in chordally ricocheting patterns all around the inner net's surface. I therefore saw that because every action has its reaction, it would be possible to pair all the molecules so that they would behave like two swimmers who dive into a swimming tank from opposite ends, meet in the middle, and then, employing each other's inertia, shove off from each other's feet in opposite directions.

surface of a balloon skin corresponding in pattern to the islanded ten-segrity geodesic strut-end positions and then to insert into those pockets stiff battens that cause the otherwise limp balloon bag to take spherical shape, as it would if filled with a pressured-in gas.

If we employ hydraulic pressure within the local islands of compression for dimensional stability and if we employ gas molecules between the liquid molecules for local shock-load compressibility (ergo, flexibility), we will find that our geodesic tensegrity structures will in every way have taken advantage of the same structural-strategy principles employed by nature in all her sizes of biological formulations.

Twelve Degrees of Freedom

I formed a tetrahedron of six 2-foot-long thin-walled steel tubes with an outside diameter of 1 inch, welded to four 3-inch-diameter steel balls at the tetrahedron's four corners (see Fig. 6.46). I drilled and tapped (threaded) four holes on the inside of the four corner balls. I then connected those four corner balls perpendicularly to a single 3-inch-diameter steel ball located at the center of volume of the tetrahedron, that center ball itself having four drilled and reverse tap-threaded holes. I made the connection of the center ball with the four corner balls by means of four 1/8-inch steel rods, each threaded oppositely at their re-spective two ends. Then I inserted the positive- and negative-threaded rods between the corner balls and the center ball and tightened them together by rotating the rods with a wrench to shorten the distance between the pulled-together end balls, as with turnbuckles. The center ball could not be dislocated from the tetrahedron's exact center of vol-ume. I then took a stillson wrench and found that without displacing the center ball from the exact center of the tetrahedron, I could rotate the ball mildly in six different positive and six different negative directions. To counteract these in-place rotatabilities required twelve rods in four sets of three tangential rods, with each rod's outer end independently fixed tangentially to each of the four outer corner balls of the enclosing tetrahedral frames and with each rod's inner end fastened only tangen-tially to the center ball. This produced twelve prime restraints on the center ball, which could no longer either be dislocated from the tetra-hedron's center or be locally twisted in place.

Recognizing that the center ball and all of the corner balls are them-selves complex microsystems, I discovered that the twelve restraints proved to be the always and only twelve restraints necessary to

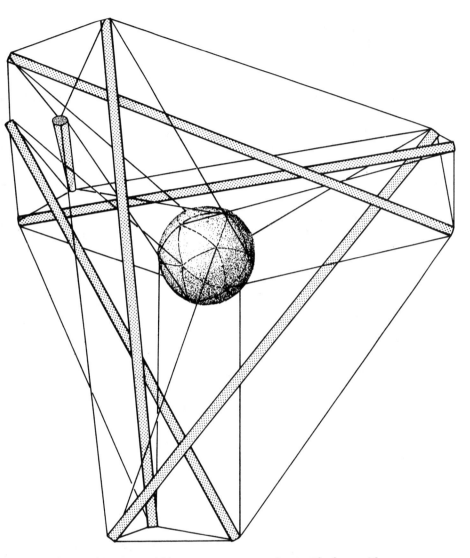

FIG. 6.46 *A system within a system: tensegrity tetrahedron with a tensionally positioned central ball suspended at its center of volume.* Central ball completely restrained in terms of all twelve degrees of freedom of all individual systems. Note that the six "solid," push-pull compression members are the acceleration vectors trying to escape from the system at either end by action and reaction, whereas both ends of each would-be escapee are restrained by four tensional wires, two long and two short, while the ball at the center is restrained from local displacement, torque, and twist by three triangulated tension wires, each also tangentially affixed to each of the four outer corner balls.

cope structurally with the twelve degrees of freedom of all independent systems in Universe. If a complex of systems is to act as one system, it is the twelve degrees of freedom and their twelve restraints that must always be structurally (push-pullingly) coped with. As previously noted, they are the same twelve restraints we found to be necessary to stabilize the wire wheel.

They disclosed the method by which the twelve degrees of freedom must be coped with to structurally associate systems within systems and to produce the interior rigidity of all superficially misidentified "solid" systems.

The four sets of three each, which all together compose the twelve-system structuring and/or intersystem structuring, are four unique additional dimensions of conventional three-dimensional phenomena: $4 \times 3 = 12 = 6$ positive $+ 6$ negative degrees of freedom.

Tensegrity Masts

The minimum structural system in Universe—the tetrahedron—can be tensegrity-structured. A linear growth of the tensegrity tetrahedron becomes a tensegrity column. Because carbon fiber is most probably constructed in exactly this way as a tensegrity tetrahedron column, it is demonstrably the strongest and lightest column structurally producible. This column and its method of assembly are shown in Fig. 6.47.

Figures 6.47, 6.48, and 6.49 show my omnitetrahedra-comprised tensegrity mast, each of whose struts in turn comprises tetrahedral tensegrity masts, each of whose micromast struts in turn consists of tetra-tensegrity masts . . . until we reach the minitude of the atoms, whose internal structuring is discontinuous compression–continuous tension. The tensegrity mast demonstrates why carbon fibers have twelve times the strength per pound of structural steel with minor carbon content and four times the strength per pound of the strongest aluminum alloy.

In 1983 Boeing Aerospace invited me to conduct a workshop on synergetics for their space station engineers. In delivering payloads into space and in other situations where weight, structural strength, and compactness are critical considerations, the principle of tensegrity will play an important role. The tensegrity mast has the additional property of being able to be delivered entirely collapsed, ready to be explosively expanded into a lightweight, structurally stable construction member upon arrival in space. This exposure of space station engineers to the principle of tensegrity could conceivably advance the space station pro-

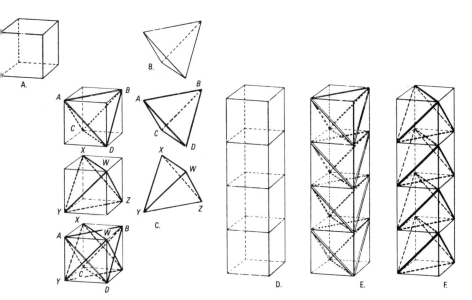

FIG. 6.47 *Functions of positive and negative tetrahedra in tensegrity stacked cubes.* Every cube has six faces (*A*). Every tetrahedron has six edges (*B*). Every cube has eight corners and every tetrahedron has four corners. Every cube contains two tetrahedra (*ABCD* and *WXYZ*) because each of its six faces has two diagonals, the positive and negative set. These may be called the symmetrically juxtaposed positive and negative tetrahedra whose centers of gravity are congruent with one another as well as congruent with the center of gravity of the cube (*C*). It is possible to stack cubes (*D*) into two columns. One column contains the positive tetrahedra (*E*), and the other contains the negative tetrahedra (*F*).

gram by many years. Following nature's own design principles, humans may be able to produce most-economic designs while at the same time solving formerly insoluble design problems.

To realize the significance of tensegrity in understanding nature's own designs and in implementing the new design science, we turn to cell biology. Don Ingber of the Yale School of Medicine, in a paper entitled "Tumor Formation and Malignant Invasion: Role of Basal Lamina," describes the role of tensegrity structure in cell architecture:

> An epithelial structure can be regarded as a tensile or tensegrity system, that is, an architectural unit of the highest efficiency which consists of *discontinuous* compression-resistant members (e.g., microtubules, cytoskeletal microfilaments, fibrillar collagen) interconnected directly or indirectly by a

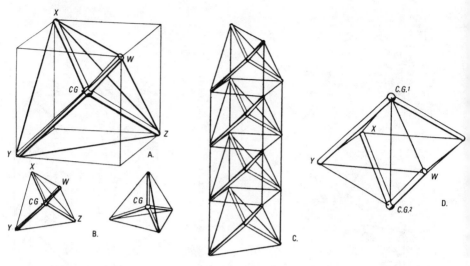

FIG. 6.48 *Stabilization of tension in tensegrity column.* We put a steel
sphere at the center of gravity of a cube which is also the center of
gravity of a tetrahedron and then run steel tubes from the center of
gravity to four corners, *W, X, Y,* and *Z,* of negative tetrahedron (*A*).
Every tetrahedron's center of gravity has four radials from the center of
gravity to the four vertexes of the tetrahedron (*B*). In the juncture
between the two tetrahedra (*D*), ball joints at the center of gravity are
pulled toward one another by a vertical tension stay, thus thrusting
universally jointed legs outward, and their outward thrust is stably
restrained by finite sling closure *WXYZ.* This system is nonredundant: a
basic discontinuous-compression continuous-tension, or "tensegrity,"
construction. It is possible to have a stack (column or mast) of center-
of-gravity radial tube tetrahedra struts (*C*) with horizontal (approximate) tension slings and vertical tension guys and diagonal tension
edges of the four superimposed tetrahedra, which, because of the (approximate) horizontal slings, cannot come any closer to one another
and, because of their vertical guys, cannot get any farther away from
one another and therefore compose a stable relationship: a structure.

continuous series of tension elements (e.g., plasma lemma, contractile
microfilaments, basal lamina). The term "tensegrity" derives from the
concept of "tensional integrity" and is a most efficient and economical
architectural system in which all loads are distributed equally over all
elements. As dynamic tensile structures, cells alter their shape until an
equilibrium configuration is attained which most efficiently and evenly
distributes the load given the characteristic architectural distribution of
anchors within the substratum. Thus, cells within a tissue might respond to
physical alterations in their environment as a coordinated unit due to the

equal and simultaneous distribution of forces to all of the elements of this organic tensegrity system.

Finally, we discover that every geometrical structure is a tensegrity. We determine that all geometrical structural systems can be encompassingly realized by only the isolated, omniislanded, discontinuous compression (repulsive) force components omniintegrated by the always-closed-back-into-itself, continuous tensional network of interattraction. This is to say that for every geometrical structural system, simple or complex, there is always a tensegrity structure (see Fig. 6.44).

Spherical Trigonometry: The Greek Sphere

As defined by the Greeks, a sphere is a surface equidistant in all directions from a point. But a surface equidistant in all directions requires the existence of the phenomenon known as "solid." Physics has found no solids, no absolute continua. An absolute continuum could have no discontinuities and ergo no beginning or ending surface. As defined, a Greek sphere could have no holes in it, since the curvature at the edges of the rims of the holes would be at differing distances from the sphere's center. Having no holes in its perfectly solid continuum, the Greek sphere could not accommodate any inbound or outbound traffic, thus being unable either to import or to export energy and ergo defying the second law of thermodynamics, by which all systems are always losing energy. It would therefore become the first local perpetually regenerative system in Universe. If that were so, the remainder of the complex, everywhere and everywhen, intertransforming, nonsimultaneous, regenerative events of Universe would be excessive and redundant. Since nature always accomplishes her events in the most economical way, she would be the solid perpetually regenerative sphere system, but no solids are in experiential evidence.

Since physics has discovered no absolutely solid continua, we find it necessary to redefine the spheric experience. Our definition of the spheric experience is "an aggregate of events approximately equidistant in approximately all directions from one small, central, minimum-system locus." "Approximately all directions" involves a vast number of measurements that would require a vast amount of time to complete, within an ever-transforming Universe that accounts for that only "approximate" equidistance. This means that the spheric experience is an aggregate of minisystem points, approximately equidistant in almost all

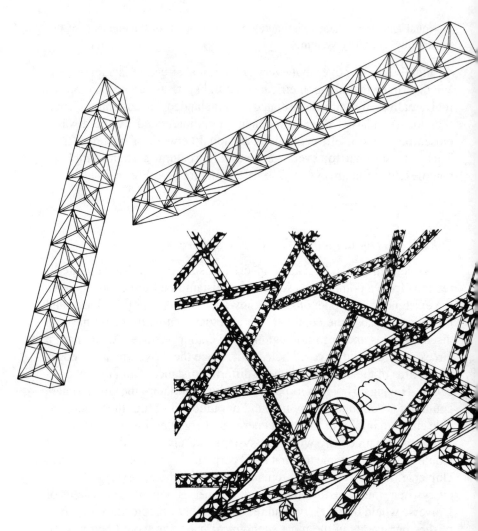

FIG. 6.49 *Tensegrity masts as struts: miniaturization approaches atomic structure.* The tensegrity masts can be substituted for the individual (so-called solid) struts in the tensegrity spheres. In each one of the separate tensegrity masts acting as struts in the tensegrity spheres, it can be seen that there are little (so-called) *solid struts*. The subminiature tensegrity mast may be substituted for each of those solid struts, and so on to sub-sub-subminiature tensegrities until we finally get down to the size of the atom, and this becomes completely compatible with the atom, for the atom is tensegrity and there are no ''solids'' left in the entire structural system. There are no solids in structures and ergo no solids in Universe. There is nothing incompatible with what we may *see* as solid at the visual level and what we are finding out to be the structural relationships in nuclear physics.

directions from one central minisystem point. Each of the spheric aggregate of points (microsystems) will have its nearest neighboring
points (systems). Most economically interconnecting those points with
their nearest neighbors involves omniintertriangulating the whole
spheric array, which means producing high-frequency geodesic
spheres in whose surface aggregation of points it will be found that
the sums of the angles around all the surface points will always be a
number that is 720° less than the total number of spheric points multiplied by 360°.

I recently made a triangle out of six stainless-steel straps, all of the
same length (18 inches). These straps were fastened together three at
each corner so that two of them make two sides of the triangle and the
third member becomes the perpendicular bisector of the triangle (see
Fig. 6.50). There are therefore three such perpendicular bisectors. The
perpendicular stainless-steel straps are made to slide by each other in the
center of the triangle. Their ends go through slots on the edge to which

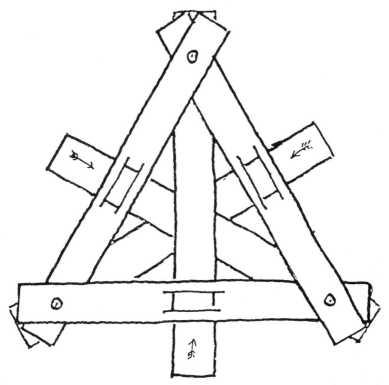

FIG. 6.50 *Model of adjustable spherical triangle made of stainless steel
straps.*

they are perpendicular. There are both in and out slots. The perpendicularly impinging ends of the stainless-steel perpendicular bisector straps jut out several inches. They can be pushed or pulled through the slot. You can slide-push the strap end inwardly through the slot. Pushing all three perpendicularly impinging ends inwardly an equal amount humps the crossing straps spherically in the middle and forces the outer triangle to go into sphericity.

There is a hole in the middle of the slot at the point where the perpendicular bisector goes through the strap. There are also three hole positions in the three perpendicularly impinging strap ends. The perpendicular bisectors cross one another at 60° angles at the triangles' center and remain in 90° perpendicularity to the edge.

When the whole triangle is flat, the three angles are 90°, 60°, and 30°, the angles of a conventional draftsman's triangle. As you push the perpendicular bisector straps inward and the triangle bows outward into a spherical triangle, at the first hole point the six right triangles read 90°, 60°, and 36° (instead of 30°). When you push the strap in further, to hole number two, the six small right triangles read 90°, 60°, 45°; in the third hole position, they read 90°, 60°, 60°.

The spherical humpings of the straps to the first hole make 90°, 60°, 36°. Twenty of these make the spherical icosahedron.

Eight of the 90°, 60°, 45° stainless-steel models make the spherical octahedron.

The model that reads 90°, 60°, 60° makes the four triangles of the spherical tetrahedron.

In the case of the spherical tetrahedron's triangles of 90°, 60°, 60°, another 30° have been added to the original 30° corner position. In the case of the spherical octahedron of 90°, 60°, 45°, the small corner triangle is 15° more than the original 30°. In the icosahedral phase, the corner triangle reading 36° is 6° more than the original 30°.

Geodesists and surveyors call these additions spherical excess. In the case of the icosahedron, there are 120 of these 6° spherical excess corners (120 × 6° = 720°).

With the spherical octahedron, where the corner is 45° instead of 30°, there is 15° of spherical excess for each corner and 48 such corners (48 × 15° = 720°).

With the spherical tetrahedron, which has 90°, 60°, and 60° in its complement of angles, or 30° more than the original, we have a total of 4 main equiangular triangles × 6, or 24 small triangles (24 × 30° = 720°).

Voilà! In each case it is 720°. This constant 720° is the sum of the angles of one regular tetrahedron. Thus, we have demonstrated that the sum of the angles around all the vertexes of any polyhedron is always evenly divisible by the number 720—that is, by one whole tetrahedron.

The sum of the angles around all the vertexes of a tetrahedron is 720°. This is true of the sum of the angles around the vertexes of any system, symmetrical or asymmetrical.

The sum of the angles around the vertexes of any system, whether it is all the outer shape-defining points of a crocodile, a giraffe, or an orange, is always evenly divisible by 720 and is always 720° less sumtotally than the numbers of outer vertexes times 360. In other words, this sum is always the remainder of subtracting one tetrahedron's 720° from the number that is the product of multiplying all the vertexes of the system by 360°.

Most important, the difference between a flat piece of paper and a polyhedron is one tetrahedron, and the difference between a polyhedron and a sphere is always one more tetrahedron, 720°. In a sphere there are always 360° around every point. This is to say that in a spherical polyhedron the sum of the angles around all its external vertexes is always one tetrahedron greater than that sum in a planar-faceted polyhedron. This means, then, that whereas the regular tetrahedron of straight edges has a volume of 1, we added one tetrahedron to make it a spherical tetrahedron, the volume of which is exactly 2.

The volume of the regular octahedron, 4, has had one tetrahedron added to it to produce its counterpart, the spherical octahedron, which has a volume of 5.

The icosahedron has a volume of 18.51 with straight edges. As a spherical icosahedron, it has a volume of 19.51, one tetrahedron added.

I was able to write out this new hierarchy of primitive systems and find it to be the initial structuring system of Universe—and so sublimely simple, with the only variables being the first four prime numbers.

UNIVERSE IS THE SUM of all positive and negative intercomplementations. To *realize* a system—for instance, a thought—means tuning in the thought and leaving all the rest of the Universe untuned. This is done by subtracting or withdrawing one tetrahedron:

$$\text{System} + \text{tetrahedron} = \text{Universe}$$

or more correctly

$$\text{System} + \text{macrotetra} + \text{microtetra} = \text{Universe}.$$

Spherical great circles are geodesics. As we recall, a geodesic is the most economical relationship between any two events. Geodesic lines are the shortest surface distances between two points on the outside of a sphere.

A great circle is that line formed on the surface of a sphere by a plane passing through the sphere's center. The Earth's equator is a great-circle geodesic; so, too, are the Earth's meridians of longitude. Any two great circles of the same system must cross each other twice in a symmetrical manner, with their crossings always 180° apart.

Now, in view of all the experimental evidence of physics, the most accurate definition of the spheric-system experience is an aggregate of energetic events approximately equidistant in approximately all directions from one approximately immobile event center. Since great circles prove to be the shortest distance between any two points on a sphere, and since the chords of spheres are shorter than the arcs of great circles, the shortest distance between any two spheric surface "events" is the great-circle chord. Also, since every surface event always has two nearest event neighbors, all the spheric experience systems may be intertriangulated; ergo, they demonstrate high-frequency spheric-cord division.

All the atoms in the surface of a highly polished steel-alloy ball bearing may be chordally intertriangulated. Circles have always been assumed to be the line formed by a plane cutting through a sphere, and a great circle has been assumed to be the line formed on the surface of a sphere by a plane passing through the exact center of a sphere, all of which required instantaneous (in no time) interacting and measuring. We have now to assume that what has always been thought of as a circle is an always finite polygon of chordal interlinkages. This fact forever banishes Newton's and Leibniz's theories positing the existence of "fluxions," and with those theories goes the familiar school textbook staple, pi. In reality we have, in their stead, only vastly high-frequency, omnichordally triangulated geodesic polyhedra.

We need never again wonder how nature uses the unwieldy and unresolvable pi (3.14159265 . . .) in calculating the construction of each of the spherical bubbles in a speedboat's wake, speculating at what point nature rounds off that unresolvable number. She does not. Computers recently have been able, with much effort, to calculate pi to the millionth-plus decimal point. Nature does not employ such uneconomical means in her design strategies, only twentieth-century scientists and high school math departments. Nature does not use unresolvable numbers in her designs.

As we have demonstrated, the sum of the angles around all the vertexes of any and all systems is always a number evenly divisible by 720° and is always a number 720° less than the number of vertexes of the system multiplied by 360°. This latter condition has been heretofore assumed to be valid—i.e., that for an infinitesimal moment, a sphere tangent to a plane is congruent with that plane, and likewise, a straight line tangent to a circle is for that same infinitesimal moment congruent with that circle. I am therefore continually seeking ways to describe the vanishment of pi, which is the misassumption that we could have absolute planar 360° surroundment of a sphere.

Since pi cannot be mathematically resolved, nature cannot use it, and you and all of us had best stop doing so or we will sacrifice our divine gift of mind, which deals exclusively with the truth.

In geodesics, it is through the strategy of using great-circle chords and not arcs that I have succeeded in triangling the sphere.

Unity is plural and at minimum two. A triangle must be bounded by something, there being no infinite planes.

The Greeks defined a triangle as an area bounded by a closed line of three edges and three angles. A triangle drawn on the Earth's surface is actually a spherical triangle described by three great-circle arcs. It is evident that the arcs divide the surface of the sphere into two areas, each of which is bounded by a closed line consisting of three edges and three angles; thus, the total area of the sphere is divided into two complementary triangles. The area apparently outside one triangle is seen to be inside the other. Because every spherical surface has two aspects—convex if viewed from outside, concave if viewed from within—each of these triangles is in itself two triangles (Fig. 6.52) Thus, one triangle becomes four when the total complex is occultly (as in astronomical convention) understood. *Drawing* or *scribing* are operational terms. It is impossible to draw without an object upon which to draw. The drawing may be made either by depositing on or by carving away—that is, by creating either a trajectory or a tracery of the operational event. All the objects upon which drawing may be operationally accomplished are structural systems having insideness and outsideness. The drawn-upon object may be symmetrical or asymmetrical, a piece of paper, a clay tablet, the surface of the Earth, or a blackboard system having insideness and outsideness.

Having now determined that a physical sphere is a closest-to-one-another assemblage of atoms equidistantly arrayed around and from a common center and, further, that the closest-to-one-another intersurface distancing of these atoms is by their chords and not their arcs, we see the nuclei of a physical sphere interpattern as an aggregate of edge-

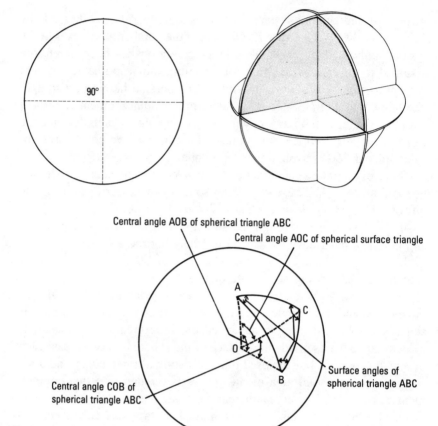

FIG. 6.51 *The spherical triangle.* The sum of the angles of a triangle is never 180°.

congruent triangles. Since the sum of the angles around the outer vertexes of these triangles is always a number less than 360°, the old concept of a plane and a sphere being for a moment the same 360° is invalid and not physically demonstrable. Atomic physics' geometry, we may therefore conclude, is non-Euclidean. These implementations of synergetic geometry have brought me to the point where I am able to say conclusively that I am beginning to comprehend incisively the structure of matter in all of its variable states, and molecular, atomic, and subatomic patterning.

We next discover that the higher the frequency of spherical tensegrity structure, the shorter the islanded compressional chords, indicating that at very high frequency the chordal struts contract to become is-

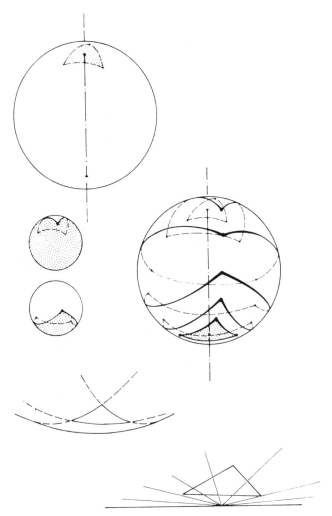

FIG. 6.52 *Triangles on surface of sphere, several views.*

landed spheres—spheres of compression. Any axis of a sphere is a neutral axis, and the high-frequency asymmetric polyhedra (the so-called spheres) contain the most volume with the least surface. The "sphere" is the unattainable limit condition of line contraction.

We then discover what has for ages disturbed physicists: the seemingly contradictory coexistence of particle discontinuity and wave continuity.

Particle discontinuity is islanded compression of Universe, and wave continuity is tensional, gravitational integrity of Universe.

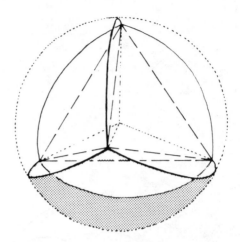

FIG. 6.53 *The four great circles of a sphere.* The spherical tetrahedron divides area of sphere into four triangulated areas (base *X* altitude), eliminating need for pi.

We have come to call this discontinuous compression with continuous tension *tensegrity.* As I described before, I coined the term to represent the universal phenomenon of tensional integrity. In tensegrity, all the system's tension vectors are inherently wavilinear and vibratible, and they always distribute their closed-system, tension-imposed stressing absolutely evenly (as the pneumatic tires distribute their internal pressures evenly to all their tensionally enclosing, high-tension-resistant tire casings).

Each tensegrity system can be overall, evenly tunable, tightened or loosened by the microcosmic and macrocosmic forces internally and externally affecting the system by its cosmic environment neighboring system.

Closed-back-on-itself continuous tension is wave; spherical islands of compression are icosahedral aggregates of tetrahedral particles. Only in an ultra-high-frequency polyvertexial system (the quasisphere) is every axis a neutral axis. Spheres are the limit-reaction conformation of all omniinterrepulsive forces. Spheres may be implosive or explosive, energy importers or exporters, planets or stars, atomic nuclei or icosahedral aggregates of tetrahedral photons.

All structural systems can be demonstrated as tensegrity models. The relative lengths of either the interpulsing or interattracting vectorial components of any and all structural systems can be determined swiftly by spherical trigonometry and slowly by *XYZ* coordinate calculus. The tension and compression components are all chords of central angles of

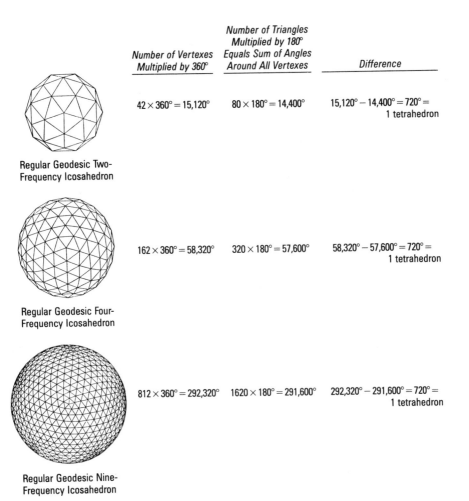

	Number of Vertexes Multiplied by 360°	Number of Triangles Multiplied by 180° Equals Sum of Angles Around All Vertexes	Difference
Regular Geodesic Two-Frequency Icosahedron	42 × 360° = 15,120°	80 × 180° = 14,400°	15,120° − 14,400° = 720° = 1 tetrahedron
Regular Geodesic Four-Frequency Icosahedron	162 × 360° = 58,320°	320 × 180° = 57,600°	58,320° − 57,600° = 720° = 1 tetrahedron
Regular Geodesic Nine-Frequency Icosahedron	812 × 360° = 292,320°	1620 × 180° = 291,600°	292,320° − 291,600° = 720° = 1 tetrahedron

FIG. 6.54 *Tetrahedral mensuration applied to spheres.*

the convergent-divergent, spherical configurations of one or more of the seven sets of unique great-circle symmetries corresponding indirectly to all seven of the crystallographic symmetries. See *Synergetics* for all such data.

The spherical trigonometry is relatively simple, and the readily available trigonometric pocket computers make it possible to obtain in minutes the chord data for any structural system you choose. If you want to use the conventional *XYZ* coordinate system, you will have to use academic science's calculus, which will take you much longer—years.

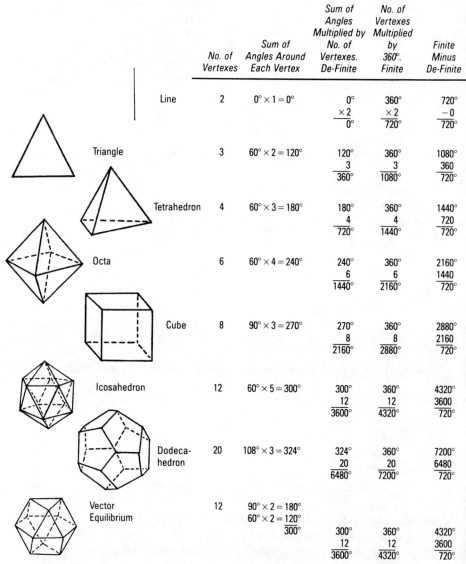

	No. of Vertexes	Sum of Angles Around Each Vertex	Sum of Angles Multiplied by No. of Vertexes. De-Finite	No. of Vertexes Multiplied by 360°. Finite	Finite Minus De-Finite
Line	2	$0° \times 1 = 0°$	$\begin{matrix}0°\\ \times 2\\ \hline 0°\end{matrix}$	$\begin{matrix}360°\\ \times 2\\ \hline 720°\end{matrix}$	$\begin{matrix}720°\\ -0\\ \hline 720°\end{matrix}$
Triangle	3	$60° \times 2 = 120°$	$\begin{matrix}120°\\ 3\\ \hline 360°\end{matrix}$	$\begin{matrix}360°\\ 3\\ \hline 1080°\end{matrix}$	$\begin{matrix}1080°\\ 360\\ \hline 720°\end{matrix}$
Tetrahedron	4	$60° \times 3 = 180°$	$\begin{matrix}180°\\ 4\\ \hline 720°\end{matrix}$	$\begin{matrix}360°\\ 4\\ \hline 1440°\end{matrix}$	$\begin{matrix}1440°\\ 720\\ \hline 720°\end{matrix}$
Octa	6	$60° \times 4 = 240°$	$\begin{matrix}240°\\ 6\\ \hline 1440°\end{matrix}$	$\begin{matrix}360°\\ 6\\ \hline 2160°\end{matrix}$	$\begin{matrix}2160°\\ 1440\\ \hline 720°\end{matrix}$
Cube	8	$90° \times 3 = 270°$	$\begin{matrix}270°\\ 8\\ \hline 2160°\end{matrix}$	$\begin{matrix}360°\\ 8\\ \hline 2880°\end{matrix}$	$\begin{matrix}2880°\\ 2160\\ \hline 720°\end{matrix}$
Icosahedron	12	$60° \times 5 = 300°$	$\begin{matrix}300°\\ 12\\ \hline 3600°\end{matrix}$	$\begin{matrix}360°\\ 12\\ \hline 4320°\end{matrix}$	$\begin{matrix}4320°\\ 3600\\ \hline 720°\end{matrix}$
Dodeca-hedron	20	$108° \times 3 = 324°$	$\begin{matrix}324°\\ 20\\ \hline 6480°\end{matrix}$	$\begin{matrix}360°\\ 20\\ \hline 7200°\end{matrix}$	$\begin{matrix}7200°\\ 6480\\ \hline 720°\end{matrix}$
Vector Equilibrium	12	$\begin{matrix}90° \times 2 = 180°\\ 60° \times 2 = 120°\\ \hline 300°\end{matrix}$	$\begin{matrix}300°\\ 12\\ \hline 3600°\end{matrix}$	$\begin{matrix}360°\\ 12\\ \hline 4320°\end{matrix}$	$\begin{matrix}4320°\\ 3600\\ \hline 720°\end{matrix}$

FIG. 6.55 *Angular topology independent of size.* Equation of angular topology:

$S + 720° = 360° X^n$, where S = the sum of all the angles around all the vertexes (crossings) and X^n = the total number of vertexes (crossings).

Tetrahedron	720°	$\frac{720°}{720°} = 1$ tetrahedron
Octahedron	240° × 6 = 1440°	$\frac{1440°}{720°} = 2$ tetrahedra
Prism	240° × 6 = 1440°	$\frac{1440°}{720°} = 2$ tetrahedra
Cube	270° × 8 = 2160°	$\frac{2160°}{720°} = 3$ tetrahedra
Icosahedron	300° × 12 = 3600°	$\frac{3600°}{720°} = 3$ tetrahedra
Rhombic Dodecahedron	109°28' × 24 = 2628° 70°32' × 24 = 1692° 2628° × 1692° = 4320°	$\frac{4320°}{720°} = 6$ tetrahedra
Dodecahedron	324° × 20 = 6480°	$\frac{6480°}{720°} = 9$ tetrahedra
Triacontahedron	130° × 60 = 10,800°	$\frac{10,800°}{720°} = 15$ tetrahedra
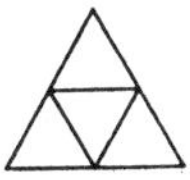 Two-Frequency Regular Geodesic	180° × 30 = 14,400°	$\frac{14,400°}{720°} = 20$ tetrahedra $= 5 \times 2^2$
Three-Frequency Alternate Geodesic	20° × 9 = 180° 180° × 180° = 32,400°	$\frac{52,400°}{720°} = 45$ tetrahedra $= 5 \times 3^2$
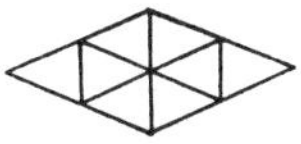 Four-Frequency Triacon Geodesic	180° × 240 = 43,200°	$\frac{43,200°}{720°} = 60$ tetrahedra $= 15 \times 2^2$

FIG. 6.56 *Tetrahedral mensuration applied to well-known polyhedra.* We discover that the sum of the angles around all vertexes of all solids is evenly divisible by the sum of the angles of a tetrahedron. The volumes of all solids may be expressed in tetrahedra.

Six Fundamental Motions of Universe: Vectors and Degrees of Freedom

There are always and everywhere insistently operative six positive and six negative degrees of freedom. All six of the degrees of freedom must be brought under local control to produce local Universe structure, which always also involves twelve comprehensively coacting, reactive, inertial complementations, which govern all such structuring. The minimum of twelve wires that hold the hub of a wire wheel stable in relation to its rim demonstrates this principle (see Fig. 6.61). Twenty-four pos-

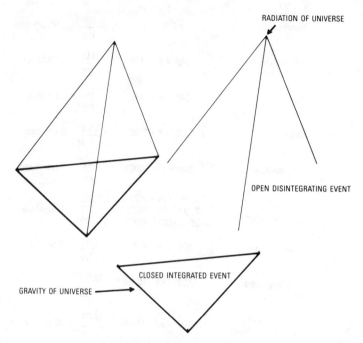

RADIATION OF UNIVERSE

OPEN DISINTEGRATING EVENT

CLOSED INTEGRATED EVENT

GRAVITY OF UNIVERSE

FIG. 6.57 *Equivector investments with opposite results.* (See also gravity radiation model, Fig. 3.3.)

itive Universe vectors and twenty-four inside-out Universe vectors are always involved.

We will go on later to discover nonunitarily conceptual Universe and its conceptual systems subdivisions of Universe in further detail, but for the moment, note that the nonsimultaneous realistic conceptualizing of the macro-, mezzo-, and micro-tune-in-able, thinkaboutable systems are characterized by electromagnetic, gravitational convergences and divergences of the local system's growths and decays, associatings and disassociatings, coexpandings and contractings. All this multiplexed convergence and divergence is inherently referenced to concentric wave-surface spheres of various radial wavelength magnitudes—all of which radii are always perpendicular to the wave-sphere surfaces and none of which radii are ever parallel to one another—and all the intercoordinating of the thinkaboutable and conceptualizable system may be realizably, definitively, and elegantly calculated in spherical trigonometry.

Spherical trigonometry's whole-system, whole-circle 360° interrelationships are alone eternally, finitely intervarying complementations of one another and are always expressible as either central angles (previ-

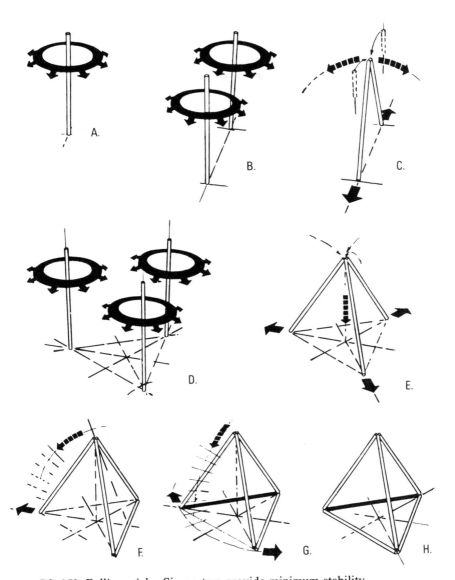

FIG. 6.58 *Falling sticks*. Six vectors provide minimum stability.

A. Stick standing alone is free to fall in any direction.

B. Two sticks: each is free to fall in any direction.

C. Two sticks: top-joined by falling toward one another and now seen as a group; free to hinge-fall and to slide apart.

D. Three sticks: free to fall in any direction.

E. Three sticks top-pointed by falling toward one another; free to have its three feet slide apart at bases and its tip ends intertwist.

F. Four sticks: a propped-up triangle, in which both the base of the triangle and the feet of the props are free to slide out.

G. Five sticks (members): two triangles may hinge outwardly and collapse as their bases hinge-slide apart.

H. Six sticks (members): complete multidimensional stability—the tetrahedron—the minimum structural system of Universe.

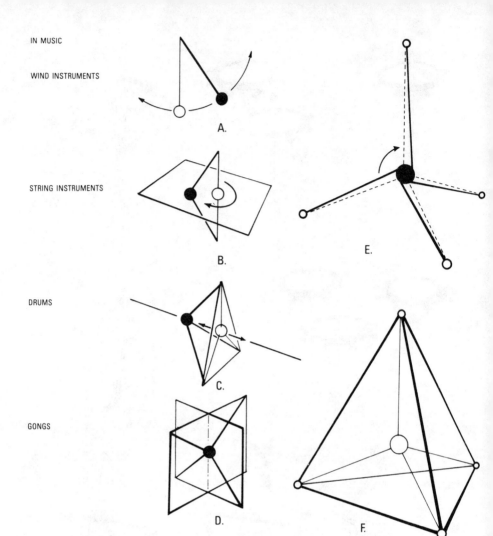

FIG. 6.59 *Four vectors of restraint define minimum system.* Music: wind instruments, string instruments, drums, gongs. Exclusively tensional investigation of the means of providing a minimum weight, structurally stable system.

A. A wavi-surfaced, varyingly radiused spheric system. Inherently the exclusively tensional restraint accommodates a constantly varying but greatest-limit radius sphere—a quasi-three-dimensional system.
B. Two tension vectors inherently define only a plane—a quasi-two-dimensional system.
C. Three tensional vectors inherently define only a line—a quasi-one-dimensional system.
D. Four tensional vectors inherently define only a point with no spatial displacement—a quasi-subdimensional system.
E. Note the possibility of in-place rotating with the position otherwise fixed by the four vectors of spatial displacement—a quasi-sub-subdimensional system.
F. The four internal tensional vectors define a physically realized structural system.

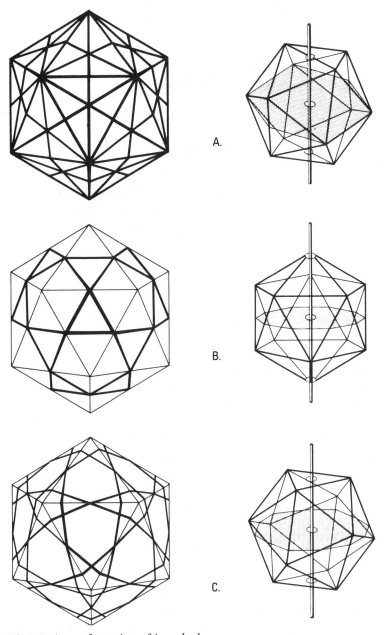

FIG. 6.60 *Axes of rotation of icosahedron.*

A. The rotation of the icosahedron on axes through midpoints of opposite edges define fifteen great-circle planes.
B. The rotation of the icosahedron on axes through opposite vertexes defines six equatorial great-circle planes, none of which pass through any vertexes.
C. The rotation of the icosahedron on axes through the centers of opposite faces defines ten equatorial great-circle planes.

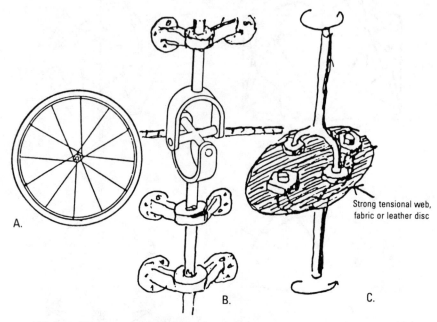

FIG. 6.61 *Minimum of twelve spokes oppose torque.* Universal joint. All the above may be considered to be tensegrity systems.

A. It takes a minimum of twelve spokes to overcome the in-place rotatabilities, despite the minimum four vectors of within-system positional restraint. This is demonstrated by the twelve-spoke wire wheel with its six positive diaphragm actions and six negative diaphragm actions, of which, respectively, three positively and three negatively oppose turbining or torquing of members.

B. Two-axis "universal joint," analogous to the wire wheel, in basic principle relies on the independent differentiation of tension and compression for its effectiveness.

C. A strong tensional web, fabric, rubber, or leather disk may serve as a continuous tensional sheet between the opposed turbining or torque members.

ously misidentified as edges of surface angles) or as the surface-angle magnitudes themselves. To spherical trigonometry, synergetics and geodesics introduces the elegantly finite closed-system frequency of modular subdividing of its component parts as governed entirely by the trigonometric relationships within one of the spherical icosahedron's 120 basic right triangles,[5] as well as within only one of the octahedron's

---

[5] Basic triangle—the lowest common denominator of spherical trigonometry, a modular subdivision of a sphere into identical spherical triangles.

8 basic triangles and within only 1 of the spherical tetrahedron's 24 basic triangles.

The modular frequency of the system's radii can only be multiplied or additionally increased by progressive subdividing of the pre-time-size, cosmically primitive state of the omnisymmetrical primitive hierarchy of omnirational, intervolumed six-conceptual system subdivisions of the Universe: the four-vertexion, the six-vertexion, the eight-vertexion, the twelve-vertexion, the fourteen-vertexion, and the twenty-vertexion, now tuned-in for thinkable consideration as the family of eternally constant, closed, finite system subdivisions of sum-totally, nonunitarily conceptual though finite, mathematically omnirational, eternally regenerative, non-simultaneously interepisoded scenario Universe.

All systems always and only have six positive and six negative primitive motion potentials—sometimes spoken of as degrees of freedom—of which the first four are integral to the system: (1) axial rotation, (2) torque, (3) expansion-contraction, (4) inside-outing (involuting-evoluting), (5) orbital travel, and (6) precession, which is the effect of systems in motion upon other systems in motion. All six of the above have their reverse behaviors.

Inside-Outing, Involuting-Evoluting

The inside-outing transformation of a triangle is usually misidenti-fied as "left versus right," as "positive and negative," or as "existence versus annihilation" in physics (Fig. 6.62).

Of all the Platonic polyhedra, only the tetrahedron can turn inside out. There are three ways it can do so: by single-, double-, and triple-bonded routes.

Inside-outing is four-dimensional and often complex. It functions as complex intro-extroverting.

A rubber glove, with its exterior colored red and its interior green, when stripped inside-out from off the left hand as red fits the right hand as green. First, the left hand was conceptual and the right hand was nonconceptual; then the process of stripping off inside-outingly *created* the right hand. And then vice versa as the next strip-off occurs. Strip it off the right hand and there it is left again. (See Fig. 6.63.)

That is the way our Universe is. There are the visibles and the invisibles of the inside-outing simultaneity. What we call thinkable is always outside out. What we call space is just exactly as real, but it is inside out. There is no such thing as right and left.

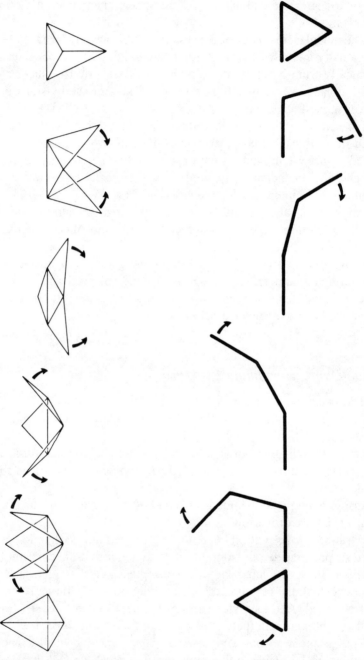

FIG. 6.62 *Implicit inside-outing of triangle*. This illustrates the inside-outing of a triangle.

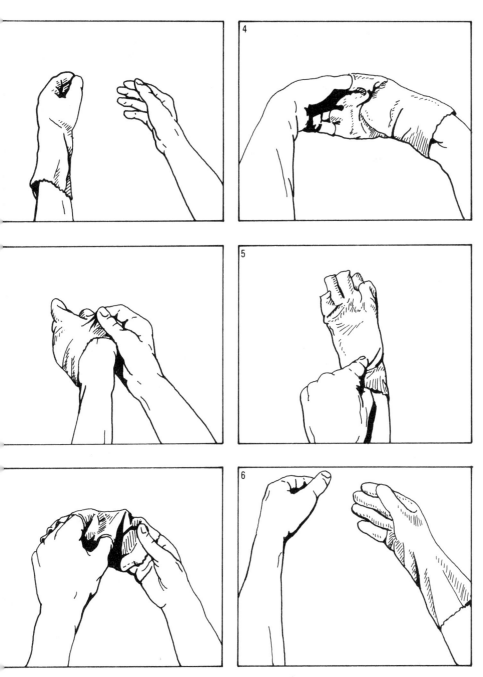

FIG. 6.63 *Inside-outing of glove.*

Orbital Travel

Of the bodies in physical Universe 99.9 percent are operating orbitally—therefore normally. As the Sun's pull on Earth produces orbiting, orbiting electrons produce directional field pulls.

The transition from being an entity to being a plurality of entities is precession, which is a peeling off into orbit rather than falling back into the original entity. Because unity is always plural and at minimum two, reality is always orbital. For the same reason, all orbits are elliptical rather than circular, having at least one additional critical proximity aberration to its very great circular orbit.

Orbit is equivalent to circuit. All terrestrial critical paths orbit the Sun. No path could possibly be linear. The Universe never reverts to the smaller, simpler circuits. (See Fig. 6.64.)

Involution and Evolution

In four-dimensional conversion from convergence to divergence, and vice versa, the terminal condition reverses evolution into involution, and vice versa. Involution occurs at the system limits of expansive intertransformability. Evolution occurs at the convergent limits of system contraction.

If we mount rubber tires on the eight triangular faces of the vector equilibrium with each tire touching other tires at three points, as in Fig. 6.65, the whole assembly can operate like a rubber doughnut. It could

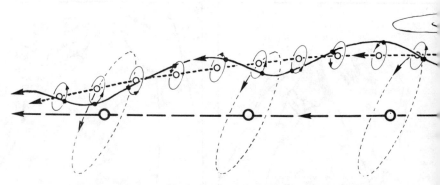

FIG. 6.64 *Reality is spiro-orbital.* All terrestrial critical path developments inherently orbit the Sun. No path can be linear. All paths are precessionally modulated by remotely operative forces producing spiralinear paths.

be rotated inward like a torus, or it could be rotated outward like an atomic-bomb mushroom cloud, coming in at the bottom and opening outward and upward at the center. Seen in their sky-returning functioning as recirculators of water, trees have an ecological patterning that is very much like a slow-motion tornado: an evoluting-involuting pattern fountaining into the sky, while the roots reverse-fountain, reaching outward, downward, and inward into the Earth again once more to recirculate and once more again—like the pattern of an atomic-bomb's cloud or electromagnetic lines of force. Fig. 6.66 shows examples of involution-evolution.

Precession

The sixth motion is precession, which we covered in some detail in the early part of this book.

To reiterate briefly, physics has two kinds of acceleration: angular and linear. When you tie a weight on the end of a string and, holding it high, rotate it around above your head, the more muscle and speed you work into it, the farther it will travel when you let go of it. That is what physics calls angular acceleration. In angular acceleration you can accumulate the energy put into the acceleration. An Olympic hammer thrower accumulates his muscle-expended energy in the circular acceleration of the steel ball on the end of his steel rod. The amount of energy

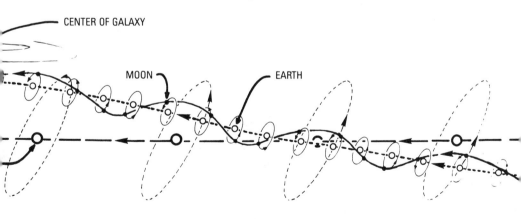

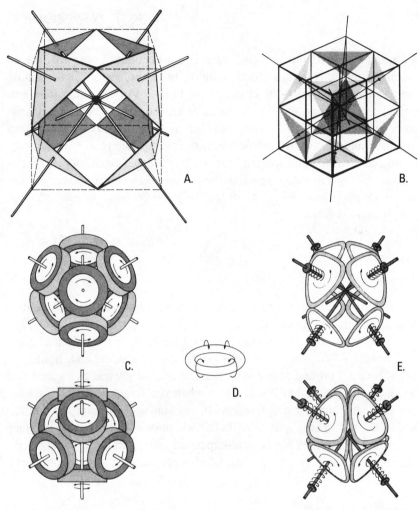

FIG. 6.65 *Four axes of vector equilibrium with rotating wheels or triangular cams.*

A. The four axes of the vector equilibrium suggesting a four-dimensional system. In the contraction of the "jitterbug" from VE to octahedron, the triangles rotate about these axes.

B. Each triangle rotates in its own cube.

C. The four axes of the vector equilibrium shown with wheels replacing the triangular faces. When one wheel is turned, the others also rotate. If one wheel is immobilized and the system is rotated on the axes of this wheel, the opposite wheel remains stationary, demonstrating the system polarity.

D. Each wheel can be visualized as rotating inwardly on itself, thereby causing all other wheels to rotate in a similar fashion. Or we can hold onto the bottom of one of the wheels and turn the rest of the system around it. If we do so, we find that the top wheel polarly opposite the one we are holding also remains motionless while all the other six rotate like an involuting torus.

E. Each wheel is conceived as a cam shape. When they rotate a continuous "pumping" or reciprocating action is introduced.

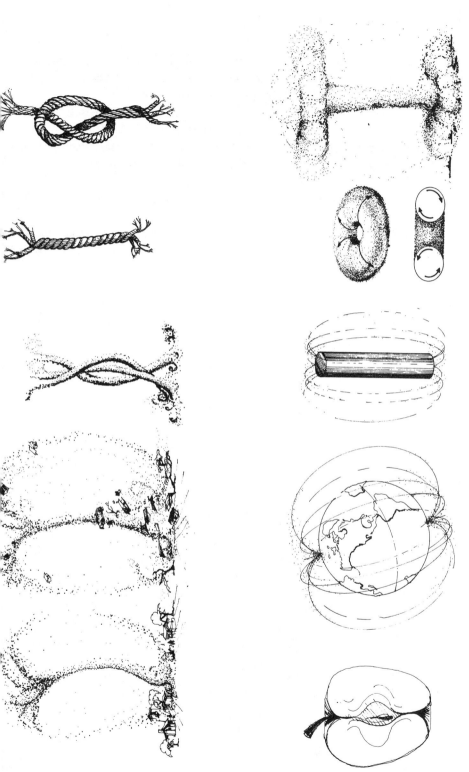

FIG. 6.66 *Involution and evolution.*

he has accumulated in the acceleration determines how far the hammer will travel when he lets go of it. The contest is to see who can accelerate the hammer so that it will fly the longest distance.

Linear acceleration is what gravity does to a body released far out from the Earth's surface—a so-called falling body. By Galileo's law, every time the Earth-approaching object halves the distance that it has yet to travel to reach the Earth's surface, the pull of gravity increases fourfold and the object's speed increases fourfold.

We are now going to describe an experiment that involves angular acceleration.

In Fig. 6.67, we have prepared a circular floor. The floor is a thick

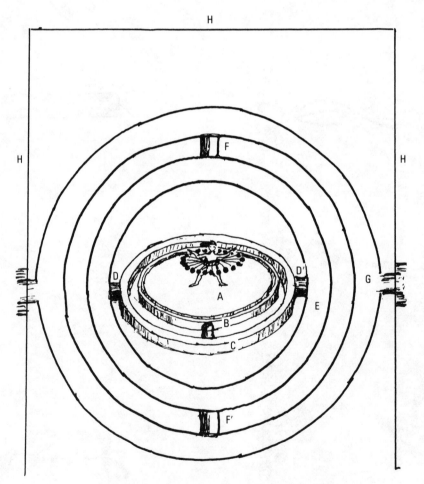

FIG. 6.67 *An experiment in angular acceleration.*

disk floated almost frictionlessly on air bearings inside the ring B, which has two 180°-apart axles turning in roller-thrust bearings mounted on the inside of ring C. Ring C itself has axles of rotation B–B' at 90° to the inner ring B's axes of rotation, which in turn is also mounted on tapered roller-thrust bearings D–D' fastened at 90° from the C ring's outer bearings on the inner side of a great aluminum annular ring E. This latter ring E is in turn roller-thrust-bearing-mounted at F–F' at points 90° from the previous axis at D–D', inside of an outermost fixed structure, ring G.

This mechanical complex of rings within rings mounted on the three 90°-to-one-another X and X', Y and Y', Z and Z' axes is what is known as a gimbal. Gyroscopes and ships' compasses are mounted in gimbals. Precession is the operative principle.

What we have described for our experiment is a giant gimbal system mounted either rotationally or fixedly inside a very large building H. We have electric switches connected to brakes on all the complex of bearings in the gimbal system. We now lock these brakes and leave that scene in building H.

Unity Is Plural and at Minimum Two

In summary, we have discovered that all geometrical structural systems can be encompassingly realized only by isolated, islanded compression units of rational (whole number) volumes integrated by a continuous-tensional network (see Fig. 6.70). Whether simple or complex, structural systems in synergetics can only be realized as whole units.

From all the foregoing we must conclude that there are no solids and, as defined, the Greek sphere could have no systemic substance or any of the topological characteristics of system. As defined by the Greeks, the sphere would have to be either an absolutely solid ball or a convex-concave shell, the inner surface of which would be of lesser radius than the outer surface.

Since you cannot demonstrate a surface of nothing, the Greek sphere could have no openings—no holes of any size. No energy could enter or exit. It would therefore defy the second law of thermodynamics, which states that all physical systems must in time lose energy. The Greek sphere would have to consist of an absolute, everywhere-undifferentiated, impenetrable, ergo inexperienceable, eternal continuum.

Since all experiences consist always and only of physically or metaphysically encountered systems and since Euler and synergetics make

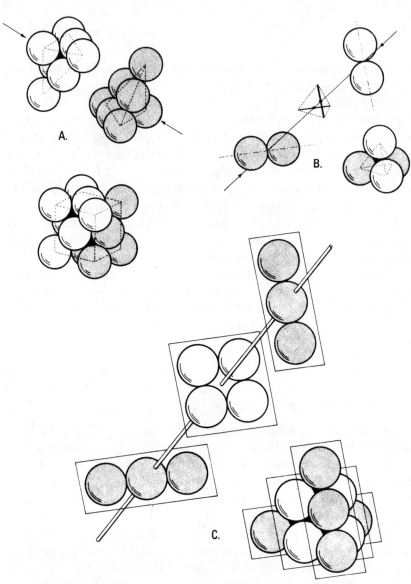

FIG. 6.68 *Tetrahedral precession of closest-packed spheres.*

A. Two pairs of seven-ball triangular sets of closest-packed spheres precess in 60° twist to associate as the cube. This fourteen-sphere cube is the minimum structural cube which may be produced by closest-packed spheres. Eight spheres will not close-pack as a cube and are utterly unstable.

B. When two sets of two tangent balls are self-interprecessed into closest packing, a half-circle interrotation effect occurs. The resulting figure is the tetrahedron.

C. The two-frequency (three-sphere-to-an-edge) square-centered tetrahedron may also be formed through one-quarter-circle precessional action.

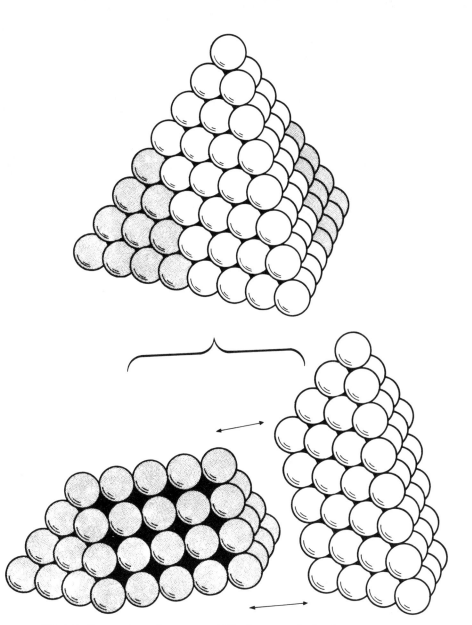

FIG. 6.69 *Precession of two sets of 60 closest-packed spheres as seven-frequency tetrahedron.* Two identical sets of 60 spheres in closest packing precess in 90° action to form a seven-frequency, eight-ball-edged tetrahedron with 120 spheres, of which exactly 100 spheres are on the surface of the tetrahedron and 20 are inside. The 120-sphere nonnucleated tetrahedron is the largest possible double-shelled tetrahedral aggregation of closest-packed spheres having no nuclear sphere.

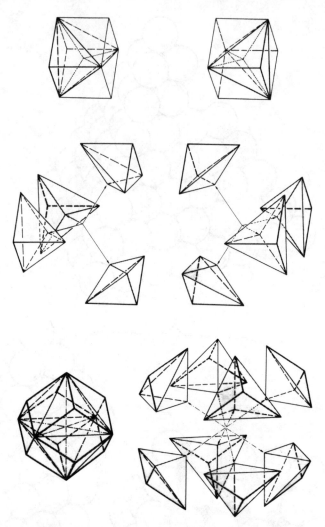

FIG. 6.70 *Rational volumes of tetrahedroning.*

A. The cube may be formed by placing four one-eighth-octahedra with their equilateral faces on the faces of a tetrahedron. Since tetrahedron volume equals 1, and one-eighth-octahedron equals ½, the volume of the cube will be $1 + 4 (½) = 3$.

B. Because there are eight one-eighth-octahedrons, with each of them equaling a half-tetrahedron, four of them can be placed on the negative tetrahedron and four on the positive tetrahedron, making a total of 2 cubes = 6, four positive quarter-tetrahedra and four negative quarter-tetrahedra superimposed on one octahedron, giving the rhombic dodecahedron a volume of 6.

C. The rhombic dodecahedron may be formed by placing eight quarter-tetrahedra with their equilateral faces on the faces of an octahedron. Since the octahedron volume equals 4, and a quarter-tetrahedron equals ¼, the volume of the rhombic dodecahedron will be $4 + 8 (¼) = 6$.

clear that all systems consist always of a minimum plurality of uniquely functioning and differentiable parts, topologically differentiable experience demonstrates that parts cannot exist separately from systems—i.e., by themselves.

There is no such phenomenon in Universe as "one," the lone observer. There is necessarily something observed. Experienceable unity is necessarily plural and at minimum two. The system's inherent insideness and outsideness, its inherent concavity of insideness aspect and inherent convexity of outsideness aspect, coexist in pure principle: one cannot exist without the other.

There is another way of demonstrating the at-minimum-twoness of the Universe (*uni-verse* means toward union, not toward isolatable oneness).

That which is concave concentrates impinging radiation, and that which is convex diffuses the same impinging radiation, so concave and convex are not the same function; ergo, the minimum otherness experience of life awareness is a system unto itself whose insideness and outsideness demonstrate that unity is always plural and at minimum two.

The other at-minimum-twoness of unity is the inside concavity and the outside convexity of the observer and the observed and their interkinetic life realization in pure principle.

Since no solids fulfill the Greek definition of a sphere as "the surface of a solid absolutely continuous in all directions from a point," we must redefine the spheric experience to that of being a closed array of separate microevents (the locus of points) approximately equidistant in approximately all directions from one approximate event atom and its complex of electrons. All those microevents at approximately equal distances in all directions from the central event will have their nearest sphericsurface neighbors occurring at the most economical (shortest) interconnection distance between them, producing a network of great-circle chords between them. Altogether this spheric array produces a closed system pattern of triangular windows—which is to say, a geodesic sphere. Polished-marble, sphere-shaped stones are, on close examination, a net of omnitriangulated windows forming a system, and that is what all geodesic domes are.

Using nature's most-economical design strategy, I first began making the tensegrity geodesic (most economically intertriangulated) domes. Keeping nature's design strategies in mind, I realized that in order to think and communicate with fidelity I now had to reidentify the number of "corners" of Euler's topology to read as the number of "somethings" and that I also had to reidentify Euler's edge lines as the number of unique structural, push-pull, vector-tensor, line-of-force "interrelationships" existing among the system's corner somethings, which interrelationships window-frame the number of different views of "nothingness" within the system.

Although some of what I came to observe, study, and explicate may seem difficult to understand, some of it is so obvious that readers may ask why it had not occurred to themselves at some time. It probably did. In childhood, spontaneous thought and unencumbered observation are quite common and simple, before the relentless disinformation process begins.

Principles are weightless. What we identify as weight is the principle of accumulative information as apprehended by the rate of our sensing data from kinetic events. The more nondirectly sensed information cognition there is, the heavier the phenomenon. In the same manner, the law of resistance of a penetrated medium by a penetrating body is that the resistance increases as N^2—i.e., as the second power of the linear speed of the penetrating body in respect to its initial resistance and as predicated upon its shape, its surface condition, and the initial viscosity of the penetrated medium. Initial resistance to a penetrating body and the seemingly inert weight of a seemingly motionless body are the same, since all is in motion and the kinetics are omni and always in operation in pure principle; ergo, *frequency* and *speed* relationships are operative only in pure principle, and principles are weightless and their local weighabilities are realizable in the pure principle of interrelativity itself.

In long-distance electric power transmission, as the voltage increases, the resistance decreases as of the third (or volumetric dimensional) power N^3, and the overall efficiency of the conducting system delivery increases as N^4. And now that we comprehend the exclusively frequency-dependent experiencing of solids, we can begin to see that weight and substance (as with so-called solids) are the consequences of (1) the magnitudes of the interrelativity timewise of linear, planar, and volume measures, and (2) the frequency in respect to abstract, weightless topology, and geometry of thinkability and its image conceptioning.

In my recently published writings, I have summarized my discovery of the option of humanity to become omnieconomically and sustainably successful on our planet while phasing out forever all use of fossil fuels and atomic energy generation other than the Sun. I have presented my plan for using our increasing technical ability to construct high-voltage, superconductive transmission lines and implement an around-the-world electrical energy grid integrating the daytime and nighttime hemispheres, thus swiftly increasing the operating capacity of the world's electrical energy system and, concomitantly, living standard in an unprecedented feat of international cooperation.

If, to the best of your knowledge and judgment, you are convinced of the technical validity of the information I submit to you, as well as of the comprehensive integrity of my commitment, I am hoping that you will study even further in my books *Critical Path* and *Synergetics,* and will commit your own genius to helping humanity understand and implement its option to use human mind for information gathering and problem solving and to apply its technological legacy to bring about peace, harmony, and an undreamed-of higher standard of living to everyone on the planet.

We are so accustomed to our school-trained linear-pattern writing, reading, and communication of information that we have failed to think spontaneously in the omniconvergent-divergent, systemic, kinetic geometry patterning of all our breathing, heart-beating, expanding-contracting, hearable-sound-and-unhearable-electromagnetic, omni-directional-wave-propagating, physical experiencing.

All living organisms grow or think "in the round," which means systemically. We expand and contract radially.

We do not live in a rectilinear, perpendicular, and parallel interpatterning of no-dimensional points, one-dimensional lines, two-dimensional planes, and three-dimensional cubes, as is still taught in all the world's schools.

Because our reflexes are academically conditioned to predominantly linear apprehending, we have failed to realize that our thoughts are inherently radially expansive and contractive, topological systems that are mathematically describable only as four- and six-dimensional systems.

General System Theory, of recent academic vintage, consists of linear lists of linearly written words on two-dimensional paper trying to describe all the linearly remembered relevant factors and parameters characterizing a given linearly experienced problem. Even "expert" parameter cerebrating at its best is mere "groping in the dark."

Laughter and loving are omniradiant, omniembracing, topologically coordinate phenomena. Love synergetically integrates metaphysical radiation and metaphysical gravity, whose interpulsative, intercomplementary oppositeness regenerates life.

The mathematical and geometrical concepts I am disclosing to you clearly comprise the rational and numerically elegant mathematical coordinate system of nature.

The history of science is replete with stories of individuals breaking free from the constraints of the conventional science of their times and initiating scientific revolutions or making great discoveries. At the root of much of this trailblazing activity is discarding in its entirety the conventional wisdom and getting to the basics—universal principles, structure, the essentials. Einstein was such an individual. His thought has changed our worldview. My experience has shown me that the discovery and practice of synergetics is an operational method and tool that is without equal for today's scientific explorers. As a case in point, I shall describe some of the outstanding events in the history of organic chemistry.

In 1852 Sir Edward Frankland discovered that organic chemistry continually manifests the numbers one, two, three, and four. At about the same time, a Russian chemist named Alexander Butlerev identified the oneness as the univalent (single) bonding of atoms into molecules, the twoness as bivalent (double) bonding, the threeness as trivalent (triple) bonding, and the fourness as quadrivalent (fourfold) bonding.

Thirty-five years later, J. H. van't Hoff asserted that the oneness, twoness, threeness, and fourness manifest by the quantitative results of the invisible behaviors of organic chemistry related to the tetrahedron. Van't Hoff was called a faker, an impostor of science, which at that time had concluded that nature used only equations and never geometrical models in her fundamental formulations.

Fortunately for van't Hoff, he lived to produce optical proof of the tetrahedral configuration of carbon. As a happy consequence, van't Hoff received the first Nobel Prize in chemistry.

This all occurred a century ago, yet neither elementary school nor university mathematics and physics departments seem to have heard the van't Hoff news. The tetrahedron is not included in any of their curricula. In its history of philosophy, the academy briefly mentions the tetrahedron as one of Plato's "solids."

Despite its universality and elegant economy, the tetrahedron has been all but ignored on planet Earth. Academic science references all its physical mensuration to the XYZ-three-dimensional coordinate system and all of its energetic phenomena to the c-g-s system, which represents the amount of energy required to lift 1 cubic centimeter of water of a given temperature 1 centimeter in 1 second of time. The cube is the chosen geometrical unit of volume measure, and the square is the geometrical unit of areal measure in all of today's world-around, state-of-the-art scientific activity, not to mention everyday use.

If you visit the General Electric Laboratories in Schenectady, New York, and witness their manufacturing of diamonds (atomically real but called artificial), you will see synthetic diamonds produced by compressing carbon to adequate degree in a powerful convergent press. The product is a complex of octahedral and tetrahedral gems of varying sizes, all in a state of intercomplementary, allspace-filling compaction.

Science opens its treatises on quantum geometry with a nonstructurally demonstrable (i.e., nonstably patterned) cube and its successive, crudely asymmetric, untriangulated fractionations. These procedures are structurally unsound, as can be demonstrated.

Take twelve rigid push-pull struts—for instance, 12-foot-long, ½-inch-diameter wooden dowels. Drill small holes through them ¼ inch from each of their ends. Take a fine Dacron fishing line and tie their ends together in groups of three. If you elevate the top two members of this assembly and hold them parallel to one another, the assembly will hang from your hands in a pseudoform, a wriggly cube. It is not triangulated and is therefore nonstructural.

If you let go of the assembly, it will immediately plop to the table or floor and collapse in a noncubical heap (see Fig. 6.71).

If you now take the same twelve struts and tie their ends together in groups of four, the whole assembly will spontaneously take its own geometrical shape, that of the octahedron (see Fig. 6.72), which will not collapse unless you apply a force greater than the tensile strength of the fishing line or greater than the compressive, buckle-resisting strength of the wooden struts. The octahedral shape persists eternally in pure principle as an omnitriangulated structure. The octahedron is eternally, inherently noncollapsible.

We use the word *primitive* to identify brain-imaginable systems whose principal structural constituents (components) are conceptual independent of size.

A tetrahedron and the four corner convergences of its six structural lines outlining four triangles is a conceptual system independent of size. Size always takes time to measure. The tetrahedron and the octahedron are primitive, pretime and presize conceptualities.

There is no such thing as a primitive cube, because it is impossible to find any position in which the three edges convergent at each of eight corners will interstabilize themselves at an omni-90° position. The way in which human society became academically hooked on the cube was by carving out rectilinearly dimensioned wall building blocks of marble while misassuming an inherent solidness to be demonstrated by the marble. We know today that there are no solids. Democritus' atoms disintegrated Plato's "solids," but proof of that waited upon Fermi's nuclear pile.

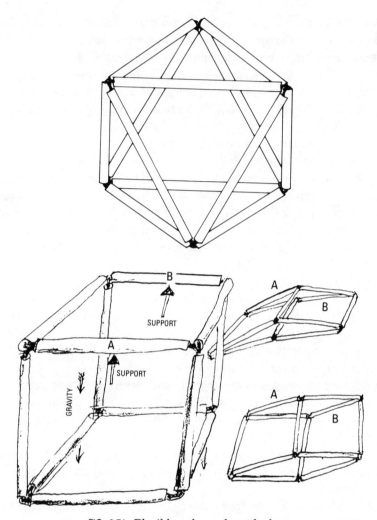

FIG. 6.71 *Flexible cube and octahedron.*

If you take six wooden struts 16.97 inches long and tie their ends into the three-together opposite corners of each of the six four-sided openings of a quasi-cubic model composed of 12-inch edges with three long struts at each corner (see Fig. 6.73). This structure is designed in such a way as to produce six struts coming together at every other corner of your eight-corner assembly. You will then have a spontaneously rigid, omnitriangulated, geometrical form that is an overall tetrahedron, which is the minimum inherently self-stabilizing system of Universe.

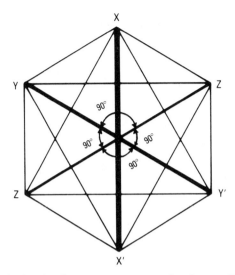

FIG. 6.72 *Octahedron's three axes cross each other at 90° at octahedron's center.*

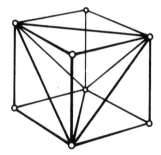

FIG. 6.73 *Cube stabilized with tetrahedron.*

In the conventional geometry mensuration taught in schools, which is based on the edge of the square and the square and the cube as the conventional modules of unity of length, area, and volume, we have, as stated in the 1982 edition of *Marks Statistical Handbook for Mechanical Engineers,* the following basic data:

With A = area of surface of polyhedra of equal edge length

V = volume of polyhedra of equal edge length

a = common edge length

	A/a^2	V/a^3
Tetrahedron, 4 triangles	1.7421	0.1179
Cube, 6 squares	6.0000	1.0000
Octahedron, 8 triangles	3.4641	.4714
Dodecahedron, 12 pentagons	20.6457	7.6631
Icosahedron, 20 triangles	8.6603	2.1813

The volume of the conventional cube is to the volume of the synergetics vector diagonal cube 1 : .9428.

Therefore, when the square and the cube are employed as unity, only the square and the cube have whole rational number areas and volumes. When the edge of the regular tetrahedron is employed as unity, the regular primitive structural systems have whole number areas and volumes.

For conversion of conventional to synergetic, omnirational-valued tetrahedral math, here are the linear, areal, and volumetric conversion factors:

Dymaxion Constants

Linear conversion factor	1.0198255
Areal conversion factor	1.0400440504
Volumetric conversion factor	1.0606605

	Area	Volume	Edge
Tetrahedron	4	1	1
Octahedron	8	4	1
Cube	1.01387	3	1.414214
Rhombic Dodecahedron		6	
Vector Equilibrium		2½ or 20	1

A plane can be defined only as a triangle. A square is always and only two equisized 90° isosceles triangles hinged together along their congruent, unit-length hypotenuses and hinged open with the two triangular planes arrayed at 180° to one another. Squarings are always $2N^2$.

As Fig. 6.74 shows, the second-powering of any number (i.e., N^2) can be experimentally demonstrated to be the number of uniformly dimensioned triangles equally subdividing the enclosed area of any triangle of the same or different-length edges, each edge being uniformly subdivided into N lengths.

All academic mathematics and all the sciences now identify N^2 as

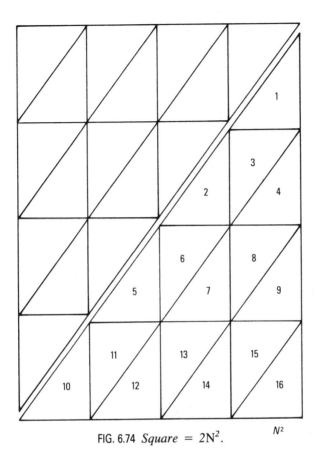

FIG. 6.74 *Square* $= 2N^2$.

N^2

"squaring." Squares or four-flex-cornered polygons will not hold their shapes (i.e., are nonstructural), and the only structurally demonstrable "square" is produced by two hinged-together-at-180° triangles.

Always operating most economically, nature second-powers the frequency of uniform subdivisions of the edges of its polygonal systems to arrive at the number of uniform subdivisions of each of the facets of any polyhedral system. If a polyhedral system has facets other than triangular, nature subdivides them into triangles to arrive at their structural stability. If you want to do your own topological accounting, you too will have to omnitriangulate, and ergo structurally stabilize, your earnestly considered polygon.

It is scientifically demonstrable that nature must always be triangling and not squaring.

Bisecting the edges of any planar figure with four different-lengthed edges (as in quadrangular accounting in Fig. 6.75) and interconnecting the bisecting points does not produce modularly dimensioned, similar four-edged figures. Bisecting the edges of any triangle, whether regular, isosceles, or scalene, always subdivides the big triangle into four always modularly dimensioned, similar triangles.

Any nonequiedged cube or hexahedron *ABCDEFGH* whose twelve edges are each divided into uniform fractional lengths, with each edge halved to start with, and that has those modular interval points interconnected with straight lines, will not be volumetrically subdivided into eight equivolumed and identical hexahedra.

Whereas any nonequiedged tetrahedron with its nonequiedges subdivided respectively into equilength linear increments, halves to start with, will always be volumetrically divided into identically volumed tetrahedra and octahedra whose octahedral volumes always exactly equal four times the volume of the tetrahedral components of the overall large tetrahedron. Nor can any asymmetric polyhedron other than the tetrahedron be uniformly subdivided into identical volumetric and linear module increments.

The tetrahedron uses only one-third the volumetric space of the cube and is therefore nature's most economic and universally employable volumetric unity and energy quantum unit.

Thus, the minimum structural system in Universe, the tetrahedron, with its six push-pull interstructuring relationships and its four corner somethings and their four opposite nothingnesses (windows), becomes the logically most structural-system-meaningful conceptuality. It has already been demonstrated that the tetrahedron has comprehensive cohering integrity. The energy involved in its comprehensive coherence is the energy of its total surface growth rate. This leads us to Einstein's energy equation $E = mc^2$, where m is the relative mass of the increment of energy considered as matter and c equals the linear speed of energy unfettered in a vacuum and c^2 equals the rate of growth in the system's energy radiantly expanding surface.

We find that we can say—indeed, must say—"triangling" instead of "squaring" when nature multiplies her linear dimensional units by themselves to arrive at a system's surface areas (N^2).

We find that we can say "tetrahedroning" instead of "cubing" when nature multiplies her system linear measurement to the third power (N^3) to obtain system volume.

Since squares, as shown by our necklace experiment, have no structural integrity, and since nature is always operating in the most economically effective way, and since every square is always two triangles

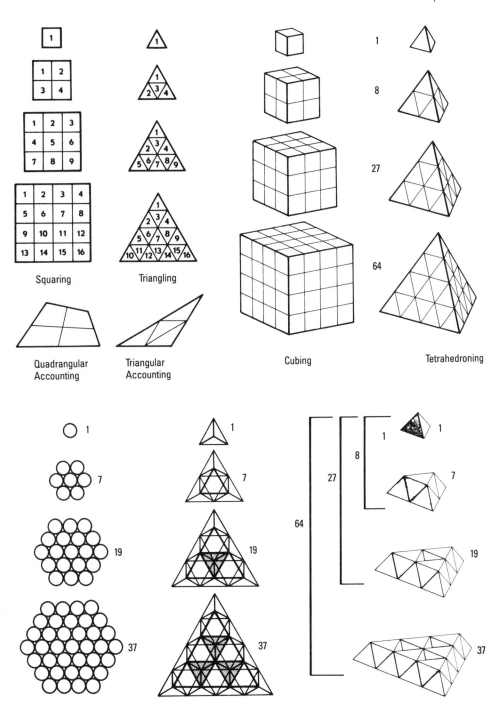

FIG. 6.75 *Quadrangular accounting, squaring and triangling, cubing and tetrahedroning.*

hinged together supposedly at 180°, and since any triangle—regular, isosceles, or asymmetrical—will do to demonstrate N^2 triangling, it is clear that nature's second-powering always and only refers to triangling.

Since necklace cubes will not hold their shape, and since the volume of two tetrahedra joined together symmetrically produces the eight vertexes of the cube whose volume is exactly three times that of the tetrahedron, and since the regular, equiedged tetrahedron is the minimum structural system of Universe and is never made asymmetrical by local asymmetrical fractionating (as shown in the cheese Platonic description below), it is obvious that nature, being most economical, must employ the tetrahedron as volumetric unity in all of her primitive systemic formulating as well as in all of her size-time interactions and intertransformings. Since the use of the cube as unity employs three times as much volume as exists in Universe, physics has to employ imaginary complex number calculations and must employ Planck's Constant of 6.625 to unburden itself of the two-thirds superfluous volume inherent in the *XYZ,* c-g-s calculations.

If we take a symmetrical polyhedron, such as a cube made of cheese, and slice parallel to one of its faces, what is left over is no longer symmetrical; it is no longer a cube. Slice one face of a cheese octahedron, and what is left over is no longer symmetrical; it is no longer an octahedron. If you try slicing parallel to one of the faces of any symmetrical geometric solid (i.e., the Platonic and Archimedean solids), what is left after the parallel slice is removed is no longer the same symmetrical polyhedron—with but one exception, the tetrahedron (see Fig. 6.76).

The tetrahedron has the extraordinary capability of remaining symmetrically coordinate and entertaining fifteen pairs of completely disparate rates of change of three different classes of energy behaviors in respect to the rest of Universe without changing its size. As such, it becomes a universal joint to couple disparate actions in Universe. For this reason, we should not be surprised at all to find nature employing such a facility for moving around Universe to accommodate all kinds of local transactions, such as coordination in organic chemistry or in the metals.

The tetrahedron's symmetry, its fifteenness, its sixness, its fourness, and its threeness are all constants. Its induced motion or position displacement to accommodate alterations in the center of gravity may explain all apparent motion of Universe. The fifteenness is unique to the icosahedron and probably valves the fifteen great circles of the icosahedron.

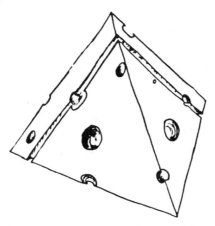

FIG. 6.76 *The cheese tetrahedron.* If you slice parallel to one of the faces of all the symmetrical geometries (i.e., all the Platonic and Archimedean "solids"), each made of cheese, what is left after the parallel slice is removed is no longer the same symmetrical polyhedron—with but one exception, the tetrahedron.

A tetrahedron is unique in its strange property of coordinate symmetry, which permits local alteration without affecting the symmetrical coordination of the whole. This means that the tetrahedron can receive changes in respect to its relation to one direction of Universe and not in respect to the other directions while at the same time maintaining its symmetry as a whole. In contradistinction to any other Platonic or Archimedean symmetrical solid, only the tetrahedron can accommodate local asymmetrical addition or subtraction without losing its cosmic symmetry. Thus, the tetrahedron becomes the only exchange agent of Universe that is not itself altered by the exchange accommodation.

There is an absolute constancy of areal, volumetric, topological, and symmetry characteristics that is exclusively unique to triangles and tetrahedra. This constancy is maintained despite any and all asymmetrical aberrations of those triangles and tetrahedra, as caused by (1) perspective distortion; (2) interproportional variations of relative lengths and angles as manifest in isosceles, scalene, acute, or obtuse system aspects; (3) truncatings parallel to triangle edges or parallel to tetrahedron faces; or (4) frequency modulations.

In contradistinction, all polygons other than the triangle and all polyhedra other than the tetrahedron exhibit a complete loss of symmetry and topological constancy as caused by any special-case, time-size alterations or changes of the perspective point from which the observations of those systems are taken.

All attempts at modeling four-dimensional cubes, the ancient Greek *tetrakytis* or the hypercube of the early twentieth century, for example, have resulted in gross distortions of size, shape, perspective, and perception. The tetrahedron (simple, quadrivalent, or unfolded as the vector equilibrium), being inherently four-dimensional, with four intercoordinate planes of mutual symmetry, undergoes no such distortion. The significance of such modeling capability becomes fully apparent in observations of four-dimensional intercoordinate rotations, such as are for the first time possible without distortion in the "jitterbug" model (see Fig. 6.77).

The fact that academic science is using the cube for unity means that physics is involving threefold the volume available in always-most-economical Universe. That is why physics must always commence analysis of the energy behavior significance of its experimentally harvested data by multiplying it by Planck's Constant, 6.625, which automatically removes the excess two-thirds volumetric value imposed by use of the c-g-s system.

The cube is structurally nonexistent in nature (except as a tertiary, nonstructural pattern aspect of the complex of vectors in an isotropic vector matrix).

Had I not started with the tetrahedron as the minimal structural system of Universe, I would not have come upon the integrity and energetic significance of the six-equivolumed A, B, S, T, and E modules.

Academia's failure to understand, acknowledge, and adopt these facts that are elementary to synergetics indicates that academia has failed altogether to understand that the omnitetrahedrally conformed A and B, S, T, and E modules would not have been discovered if I had not altogether cast out the cube from its role in present-day physics.

IN RECOGNITION OF THERE BEING NO true spheres and only high-frequency polyvertexia, when we speak and think of unit-radius spheres close packed together around the one sphere and of them being further packed together around the nuclear sphere in the always eight-triangle and six-square pattern, the sphere centers of which aggregates produce what we have shown and described elsewhere as the isotropic vector matrix, it becomes appropriate to consider what the orientation to one another of the unit-radius polyvertexia may be, since they could be symmetrically interrelated in three ways: univalent, bivalent, trivalent. They may connect one vertex to one vertex, two vertexes to two vertexes (edge to edge), and three vertexes to three vertexes (window to window, face to face). The first way (one vertex to one vertex) produces gases; the

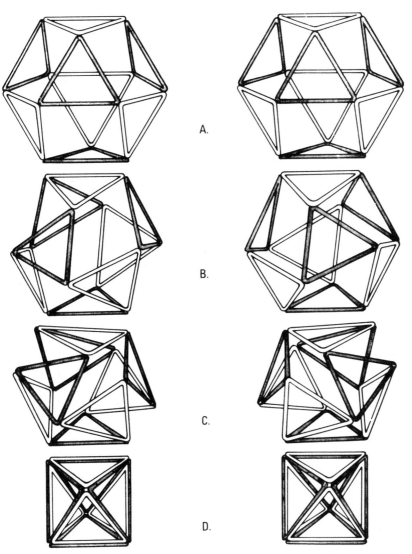

FIG. 6.77 *Symmetrical contraction of vector equilibrium: jitterbug system.*
If the vector equilibrium is constructed with circumferential vectors only
and joined with flexible connectors, it will contract symmetrically because of the instability of the square faces. This contraction is identical
to the contraction of the concentric sphere packing when its nuclear
sphere is removed. This system of transformation has been referred to as
the ''jitterbug.'' Its various phases are shown in both left- and right-hand
contraction.

A. Vector equilibrium phase: the beginning of the transformation.
B. Icosahedron phase: when the short-diagonal dimension of the
    quadrilateral face is equal to the vector-equilibrium edge length,
    twenty equilateral triangular faces are formed.
C. Octahedron phase: note the doubling of the edges.

second way (two vertexes to two vertexes) produces liquids; and the third way produces crystals (i.e., superficial "solids"). The first way (as gases) uses the greatest diameter; the second way, a lesser diameter; and the third way, the least diameter.

The least diameter produces what used to be called spheres, which, we now learn, do not touch one another when closest packed.

The isotropic vector matrix is the condition of which Avogadro speaks in which all the conditions of heat and pressure (expansion and contraction) are identical—the positive and negative vector.

What can touch one another are only gases. This gives importance to Avogadro's law that under identical conditions of pressure and heat, all gases disclose the same number of molecules per given volume. This is what we have as the interpatterning of the isotropic vector matrix, all of which proves that "solidly" speaking or crystallinely speaking, nothing in Universe touches anything else in Universe, and all is cohered only remotely by tensegrity (tensional integrity).

Since mites (quarks) are the minimum allspace fillers and since they can fill it vertex to vertex (quadrivalently), they can fill with any proportionality of positive and negative mites. That they can do so—the vacancy option—explains why ice can, and does, float on water.

Since more than one event cannot occupy the same point at the same time and since more than one event cannot passage any one point in Universe at the same time, two great circles cannot cross one another at the same radius from the system center, wherefore all of the seven foldable-into-bow-tie patterns that may be associated to seemingly reestablish their circulating of the seven systems of symmetry are demonstrating only approaches to the point of relayable continuance of their most economic travel. Their approach to points of relay can readily induce a transmitted momentum (as do hung rows of metal spheres), wherefore we may now understand that electromagnetic waves are not continuous, except in their continuum of local Lissajous figures, and that wavelike particles are finite packages.

The fact that all the seven great circles inherently fold into simple and complex bow ties, all of which are reintegratable to produce spheres, seemed at first "interesting" to me. Then it became evident that the individual 360° basic wave cycle that each manifests provides a means of holding information in a local self-regenerative shunting pattern, with releasability into cosmic travel through the tangency points inherent in closest-packed-together unit-radius spheres.

A surprising manifest of this model was that the great-circle tracking was interconnectable at the twelve tangent points only as a gap-jumping.

Such arcing may in the future explain radiant energy as a demonstration of discontinuous photon trains.

AS WE HAVE DETERMINED, LINES MAY BE more accurately described as trajectories of events.

Since two lines and their respective events cannot go through the same point at the same time, we have interferences, reflections, refractions, and smashups. This is what is discovered in the atom smashers and their peripheral cloud chambers.

A great circle is said to be a line formed on the surface of a sphere by a plane going through the center of that sphere. Great circles are said to be the shortest distance between points on a spherical surface. In spherical trigonometry, two great circles must always cross each other twice.

We are now forced to say that since lines cannot go through the same point at the same time, great circles cannot go through the same point at the same time. Great circles' tracks are not the shortest distance between two points on a sphere; the chords between those points are the shortest distances. In a polyvertexion of very high frequency, the continuum of chords may seem to go through the same point at the same time, but that cannot be. What we must conclude is that in view of the fact that two lines cannot go through the same point at the same time, all of the "foldable great circles" (which can be vertex-fastened together to seemingly reconstitute the great circles) represent the actual and only possible pathways of energy.

Lissajous figures (Bowditch curves) were discovered as useful tools for science during the early days of World War II. These figure-eight patterns were found to be the patterns that energy was producing in the microworld. I found that the seven unique cases of the foldable great circles which can be interassembled vertex to vertex seemingly reconstitute the three, four, six, and twelve great circles of the vector equilibrium and the ten, fifteen, and six great circles of the icosahedron. These represent the fundamental self-interference patterns of nature trying to achieve most economic travel routes—that is, the shortest distances between points. Energy continually recompletes these cycles. In this way, nature can have local holding patterns of energy, which can, however, be gap-jumped into wave continuums.

Synergetics, through modeling, provides this demonstration of how continuous waves and packeted quanta can be reconciled, which I shall further describe now.

All but the six great circles on the icosahedron go through the twelve main vertexes of the system. In the case of the vector equilibrium, the twelve vertexes are in even-numbered rotational symmetry, whereas the twelve vertexes of the icosahedron are in odd-numbered rotational lock. By removing the nuclear sphere of the vector equilibrium, the twelve spheres of the icosahedron collapse into the nonsymmetrical position. This could be a way of shutting off a circuit: circuits that are open on the same twelve vertexes as are open on the vector equilibrium can be cut off by collapsing the central sphere of the vector equilibrium. (See Fig. 6.78).

The twenty-five great circles on the vector equilibrium all pass through the twelve points of tangency of the spheric system with other spheric systems in closest packing. In fact, energy always follows the convex surface, which is always the most highly tensioned surface. If you bend a piece of flexible steel, the outside surface goes into higher tension, and the inside, into compression. Electrical energy always follows the highest tension. You can safely walk around inside a 20-foot copper sphere that has 2 million volts statically introduced at the surface.

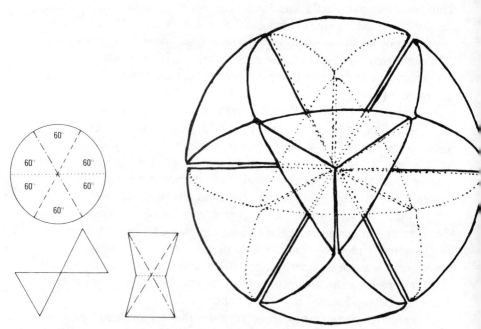

FIG. 6.78 *Vector equilibrium constructed of four foldable great circles.* As with the other polyhedra, a vector equilibrium may be constructed of great circles cut from paper.

The shortest routes through Universe would be from sphere to sphere, following those great circles that alone go through the twelve points of tangency of the spheric systems.

The foldable set of twenty-five great circles of the vector equilibrium are the only great circles that can be so folded. There are no other known cases of foldable great circles. We have only one case of nonpassage through the twelve points of tangency of the vector equilibrium and twelve points of shunted energy of the icosahedron—those being the six great circles of the icosahedron, which constitute the six equators of the icosahedron.

It is perfectly clear that at the point of contact of the folded great circle with its counterpart of the great circles, there is a true gap, which explains the phenomena—not explained by physicists—of what seem to be particles and waves. Waves seem to be continuous, and particles seem to be discontinuous. Now we can see that the wave is also the particle: it is the Lissajous figure, which with the right gap closing would constitute a circuit. Tuning, I am sure, is exactly this closing of a gap. A solenoid, for example, is used to tune to the right number of coils to allow a gap to be closed.

There is, as we have shown, no absolute continuum of anything. Higher and higher tensions are built up until one is able to "arc the gap"—to cross the gap of the (appropriately named) arc altitude between two apparently touching noncontinuous spheric systems.

From here we look at Ohm's law, which states that the amount of current equals the resistance divided by the voltage. The resistance builds up, and suddenly the charge jumps across.

WHEN I DEVELOPED MY DYMAXION MAP, *Life* magazine brought in five experts: Dr. Boggs, the chief geographer and cartographer for the U.S. State Department; the president of the American Geographical Society; and three mathematicians recommended by New York University, all of whom said that the cartographic projection I had developed was pure invention and did not conform to any known mathematics. It was easy for my patent attorney, using this information, to get the projection system granted a patent.

I told these experts that I had a three-way grid of great circles. They said that there was no such thing as a three-way grid of great circles. They overlooked the spherical octahedron, which we learn can only be done with six great circles.

My friend Mr. Norquist, who was president of Butler, the grain bin company in Kansas City, told me he could spin very accurate hemispheres. We ordered two hemispheres spun in copper $\frac{1}{16}$ inch thick, one sliding spherically around inside the other. They were precisely machined hemispheres. Their edges were great circles. They became great-circle rulers. I put half of the spherical vector equilibrium on one of them. There were four of the eight triangles, to the edges of which I inscribed perpendiculars. I started at the poles and went from pole to pole, from triangles into the squares. I got down to 60° (exactly one edge of the vector equilibrium). I marked 1° positions, and I scribed great circles from pole to pole. There was a square grid in the six spherical squares and a triangular grid in the eight triangles; this formed a three-way grid inside the triangle and a two-way grid inside the square. The 1° grid was very carefully scribed. When I began experimenting with spherical trigonometry and geodesic domes, I was convinced that there was a three-way grid of great circles within a triangle. When I began doing my spherical trig for the geodesic dome, I found that the lines of the grid did not cross exactly, they made little triangles with approximately 15 minutes of arc to the edge. It seems that what nature is doing is weaving in and out with the three-way grid.

Since you cannot go through the same point at the same time, nothing could be more wonderful or natural than these little triangles that at first annoyed me when I saw them as a discrepancy. I proved that there really could be a three-way grid, but it had to be a woven three-way grid. The weave would have to be close to the thickness of the material you were weaving.

Brain, Mind, and Universe

Elaborating on my earlier definition of Universe as "the aggregate of all humanity's consciously apprehended and communicated-to-self-or-others experiences," I note that to each individual human, Universe is the ever-multiplying totality of a uniquely evolving, special-case history of omnidimensional, omnidirectional, omnimagnitude, omnifrequency, self-and-all-others lived-in scenario.

Scenario Universe is a plurality of individual, nonsimultaneously occurring overlappings and interweavings of both unique and similar episodic characters, things, scenes, and themes.

The complex overlappings and interweavings are omnisensed, imagined, compared, and remembered only and entirely within the individual human's brain as an ever-local, time-space-conceived, evolving summary-complex of sought-for and progressively assumed-to-be personally discovered concepts of specific phenomena interrelationships holding various relative magnitudes of significance.

The relative-significance judgments by the human individual are continually translated and fed back into anticipatory reorganization of the individual's initiatives, criteria of judgments, and attitudes.

All of the foregoing subjective and objective individual prospecting and formulating always and only constitute special-case realizations of the eternally regenerative total complex of the ever nonsimultaneously integrating, intertransforming, growthfully converging and dissipatingly diverging importing-here, exporting-there sortings, selectings, combinings, implementations, and realizations of the potentials of the omniinteraccommodative cosmic totality—the family of thus-far-discovered, generalized metaphysical principles governing all the only special-case, physical realizations of Universe.

The unity of Universe is an inherently plural, complex unison. It is a uniting of all known human experience of all the special-case realizations of the eternal plurality of individually unique generalized principles.

All of the thus-far-discovered and only mathematically definable generalized principles have been discovered solely by human minds. Human minds alone are always concerned with the discovery of the inherently nonsensorial, non-brain-apprehensible interrelationships of Universe. Brains deal only in the special-case experience with temporal beginnings and endings. Minds of humans reconsidering the special-case experiencings recalled from the brain's memory banks alone are admitted to discovering and objectively employing the eternal principles of Universe.

As a consequence of the unique functioning in Universe of human mind and its discovery and objective use of the omniinteraccommodative generalized principles, as set forth in the foregoing complex statements, there has been an extraordinary harvest of significant knowledge and human capability advantaging, enumerated below.

1. The energy of the nonsimultaneously overlapping episodes of eternally regenerative Universe is only sum-totally but never omnisimultaneously constant. All energy-event multiplication in Universe is accomplished only by dividing the never-at-any-one-instant sum-totally available *energy* into progressively greater numbers of progressively more-frequent and smaller-magnitude events. Multiplication only by division ranges all the way from eternally tranquil novent,[6] through a

---

[6] There is no static geometry. There are only events and lack of events. My contraction for the limit condition of no events is "novent."

few infrequent macrocataclysmic events (e.g., novae), to many frequent microminitude events (e.g., microbes). This multiplication only by division of the total energy of Universe is uniquely identified with quantum mechanics.

2. Galileo's law of similitude is manifest in the succession of relative magnitudes of energies involved in iceberg melting, during which process there is an initial slowness of melting, because an iceberg melts only as it takes in energy as heat from outside through its relatively small surface area—as it is proportioned numerically to its volumetric mass. However, its volume becomes progressively smaller at a velocity of N^3, while its surface gets smaller only at a rate of N^2, wherefore as the iceberg melts it admits heat ever faster to melt its interior mass. We can see how its volume is decreasing at a far faster rate than the accelerating rate of admittance of outside heat which accomplishes the melting. We witness the iceberg's last frozen remainder vanish ever more rapidly.

Let us think next about Galileo's similitude law in respect to this melting and as also manifest in the case of the 18–1 slenderness-ratioed, cigar-shaped piece of steel 6 feet long that swiftly sinks into water while its 1½-inch-long steel needle counterpart floats on the same deep, still-surfaced water, wherewith we realize that both the iceberg and the steel cigar and needles manifestly demonstrate that going from the macrocosmic to the microcosmic, the volume-mass-weight relationship becomes progressively less energetically significant in respect to the now energetically great significance of the surfaces, and that surfaces in turn become progressively less significant in respect to exclusively linear interrelationships, such as those of gravity or electromagnetic proximities.

It is thus that we note the increasing interattractiveness of bodies with the diminution of the size of the bodies and their linear interdistancing. The astonishing coherence of the atomic nucleus is thus explainable, as is, to an only somewhat less dramatic degree, the ever-more-with-less tensile strength of coherence augmentations of metallic alloys.

Let us also think about the way in which this Galileo principle governs nature's own designing of all zoological creatures and botanical species—for instance, in his book *The Seven Mysteries of Life,* Guy Murchie gives the example of a mouse jumping out of an airplane and landing safely because its skin acts successfully as a parachute in arresting the mouse's rate of descent. This would not be the case with a human being, because of his greater weight per unit of skin area, or with an elephant, with its even greater weight-to-skin-area environment-imposed limitations of behavior. A human can high-jump about a foot and a half more than the human height and pole-vault about three times that height, but the fall from the latter height has to be carefully cushioned if the human legs are not to be broken. A grasshopper, on the other hand, can jump fifty times its height and land without harming itself because of its great body-surface-to-weight ratio, its jumping strength being vested in surface-tension mechanisms. Murchie cites many of these relative-size controls of the life-styles of biological life in relation to environment. His figures on hummingbird energy requirements and their rate of refueling are all part of the same mathematically stable topological, geometrical, energy-vector-quanta, chemical-bonding, electromagnetic, and gravitational relative-magnitude laws. Shipbuilders long ago learned that doubling the length of a ship increases its payload capacity eightfold but increases its hull area only fourfold—thus saving on construction cost and friction with the sea—and doubles the earning potential.

For my own part, I learned long ago that not only do spherical structures contain the most volume with the least surface but also that the curved (inherently triangulated) structure of spheres gives the greatest strength per pound of materials employed. Every time I doubled the diameter of my spheric-domical structures, I increased the contained volume of atmosphere eightfold while only fourfolding the amount of structural shell per each enclosed molecule of atmosphere, through which enclosing skin the contained molecules of atmosphere could gain or lose heat.

I have also found this same energetic-effectiveness increase as relative size is diminished to be mathematically and incisively demonstrable in going from the triple-bonded crystal's rigid structuring to the flexible viscosity of the double-bonded liquids, whose surface tension embracingly coheres a droplet of liquid or spherically embraces a bubble—whereas the only singly interbonded gas, with its atom-structured molecules, cannot maintain a system integrity very easily and to be locally retained must be enclosed within sealed containers.

In summarizing the concepts of volume, surface, and line, and quarternary, tertiary, dual, and angle bonding, the material covered earlier in this volume shows that the minimum demonstrable-reality ''something'' is a system having insideness and outsideness, and ergo the tetrahedron is the minimum considerable system; that a seemingly surface-only phenomenon is a tetrahedron of almost zero altitude; that a line is realistically a tetrahedron of great altitude and almost zero base whose altitude could be extended as long as there is time; and that its frequency within Universe is also shown by synergetics in the terms of the A Quanta Module, whose linear energetic content is constant, its wavelength always being measured from base to pinnacle.

3. First, combining (a) all the foregoing design-science considerations of the relative-magnitude and quantum-mechanics conceptioning with (b) Newton's relative-mass laws and his second-power-varying, remote-from-one-another body-interattractiveness laws and with (c) my own vector equilibrium's experimental redemonstrability of the four-dimensional jitterbug's twenty volumes omnisymmetrically and omniconcentrically contracting to one volume while intertransforming from VE to icosahedron to octahedron to tetrahedron and from single interbonding to double to triple to quadruple interbonding—i.e., to fourfold tetravertexion vectorial congruence. All of which follows a complex of elegantly statable, omnirational, mathematical-transformation laws.

We have altogether the Galilean similitude progression of volume decreasing at a third power N^3 rate while the surface of the same symmetrically shrinking geometrical form decreases at a rate of only N^2 and the linear dimensions of the symmetrically shrinking geometrical object decrease at only a first-power rate N, an arithmetical rate. This similitude progression saw the steel needle 4 inches in diameter and 72 inches long with a "slenderness" (diameter-to-length ratio of 1–18½, similar to that of Greek columns). We saw such a steel cigar sink swiftly in deep water, whereas the same steel cigar reduced to a length of 2 inches becomes a delicate steel sewing needle that floats on the same deep water, weight having become negligible and only the surface tension of the water and the surface-maintained structural integrity of the steel cigar having environmental behavior significance. We are now going to marry this Galilean similitude progression with my synergetics geometrical hierarchy's progression of volume-to-surface interrelationship changing with symmetrical intertransforming as the mass (volume/weight) changes as well as the vector structuring transforms from single to double, to triple, to fourfold, doubling up all vector lines, which alters the system's internal coherence at first-, second-, third-, and fourth-power rates as their shrinking interproximities are governed by the Newtonian mass-interattractiveness law.

In this altogether considered marriage of the similitude progression of an object system's progressively changing behavioral relationship to a given environment (e.g., the spider falling off a cliff versus the elephant falling off the same cliff), to the synergetics principles of symmetrical shape and mass transformation, we then extend these progressions to the surface by examining synergetics asymmetrical transformation in an only-altitude-decreasing transformation of a tetrahedron that approaches an almost but never entirely flat and volumeless triangular base plane. After this, we have the progression of surface contraction at a second-power rate N^2 becoming insignificant in respect to the only-arithmetical rate of shrinking the system's linear dimension as we get ever smaller. We learn in synergetics that what seems to be a line is in effect a tetrahedron whose base dimension is shrinking faster than its altitude is (see Fig. 6.79).

Then we come to synergetics' fractionation of all its hierarchy into the univolumed A, B, T, and other $\frac{1}{24}$ regular tetrahedral volume units, and to these A and B phases as constituting nature's minimal allspace filler of Universe, and to the successive quartering into ever "flatter" tetrahedra of the particles themselves. And then we come to the C, D, E, F modules to any linear extension to interreach any bodies in Universe, with all the intercohering strength of all Universe progressively concentrated to provide the intercosmic tensional capabilities discovered by Kepler to be comprehensively manifest. We also discover that at the Einstein module level, all the energies become transformed into radiation, only to have the pushed radiation bending back on itself to become eventually inward-bound as by photosynthesis it is converted into biomatter, as Murchie so magnificently discloses in *The Seven Mysteries of Life*.

Unlike other polyhedra, the tetrahedron exhibits constant properties with respect to altitude, volume, and cross-section (see Fig. 6.80).

We find as we look ever microward that bio-"life" progressively miniaturizes its componentation until it crosses the threshold between bio and crystal, after which the progression of similitude and synergetics takes over, again giving us a magnificent overview of eternally regenerative Universe.

4. Because of my physical-model-proven knowledge of the Einstein model's transdeformation in a light photon as $mc^2 = $ mass times the speed of the spherical surface growth rate of radiation expansion converted into a single tensor.

Modeling transformation and its altering noncontact, intercoherence augmentations in the following exposition of metallic alloying interaugmentation which require meshing of event patternings.

And finally with my physically proven discovery that a triangle is the only polygon that holds its shape.

5. With my discovery that there are only three primitive structural

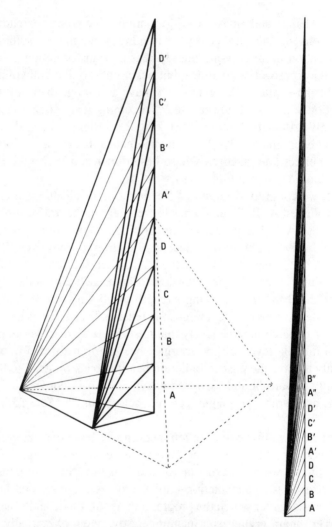

FIG. 6.79 *Constant-unit-volume progressions of asymmetric tetrahedra.*
In this progression of ever-more-asymmetric tetrahedra, only the sixth
edge remains constant. Tetrahedral wavelength and tuning permit any
two points in Universe to connect with any other two points in Universe.

systems in Universe: the six-vertexion, the four-vertexion, and the
twelve-vertexion.

6. With our proven knowledge that there always and only are twelve
degrees of freedom in Universe, six positive and six negative.

7. With my own proven knowledge of tensegrity, in which no compressional component of a structural system ever touches another.

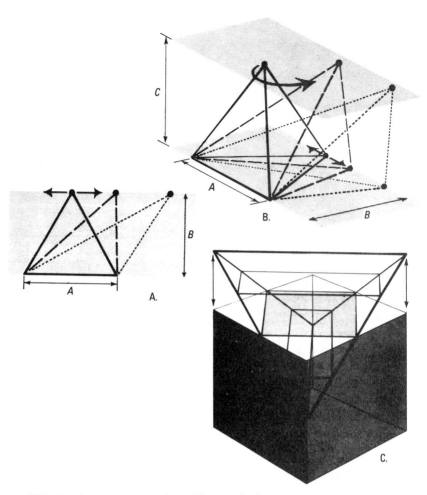

FIG. 6.80 *Constant properties of the tetrahedron.*

A. The area of a triangle is one-half the base times the altitude. Any arbitrary triangle will have the same area as any other triangle so long as they have a common base and altitude. Here is shown a system with two constants, *A* and *B*, and two variables—the edges of the triangle excepting *A*.

B. The volume of a tetrahedron is one-third the base area times the altitude. Any arbitrary tetrahedron will have a volume equal to any other tetrahedron so long as they have common base areas and common altitudes. Here is shown a system in which there are three constants, *A, B, C,* and five variables—all the tetrahedron edges excluding *A*.

C. As the tetrahedron is pulled out from the cube, the circumference around the tetrahedron remains equal when taken at the points where cube and tetrahedron edges cross; i.e., any *rectangular* plane taken through the regular tetrahedron will have a circumference equal to any other rectangular plane taken through the same tetrahedron, and this circumference will be twice the length of the tetrahedron edge.

8. With my knowledge that every seemingly solid compressional strut of a tensegrity structural system can be replaced by a tensegrity mast.

9. Because of my knowledge that a nucleated tensegrity four-vertexion requires eighteen tensional restraints externally and twelve internally.

10. Because of my knowledge of the octahedron as conservation and annihilation model, in which one unit of volume can be lost and re-gained within the same energy restraints (see *Synergetics*, 985.08; 935–38; Fig. 936.12 and Color Plate 6).

All the foregoing seems possibly to explain why physicists have been confounded by the fact that the magnitude of atomic-nucleus integrity of self-coherence greatly exceeded explainability by Newton's second-power law.

By way of example, I can explain the reason why the atomic nucleus is so dense. It is the limit condition. It is structurally quadrivalent (to the fourth power, so to speak, i.e., N^8), which can be demonstrated by the jitterbug model. Its quadrivalence is complemented by a quadrivalence of negative Universe (see Fig. 6.81).

In *Synergetics* I published the clear identity of nature's minimum allspace-filling tetravertexion, consisting of two back-to-back A mod-ules and one B module whose respective internal reflectivities' energy-releasing behaviors exactly correspond with the quarks, and went on to show that the A's and B's could be exactly quartered around their volume centers—whereby their quartering always produced tetraver-texia, and such successive quartering produced tetravertexia, ad infin-itum. All of which corresponds with the experimental results of ever more powerful ultra-ultra-high-power atom smashing.

I must conclude that the present preoccupation of the world's phys-icists is to use billions of dollars' worth of atom smashers to discover

FIG. 6.81 *Four different ways in one, i.e., four congruent tetrahedra.* This omnicongruence of atomic nuclei is also demonstrated in the chem-ical bonding of diamonds and alloying of metals.

something about the nucleus of the atom, which is akin to smashing a Boeing 747 in order to discover how its 500,000 component parts fit together in one functional design.

All that physicists need to do is study synergetics to learn how nature designed atoms and combinations of them—in pure principle.

ALL THE CELESTIALLY EVERYWHERE AND everywhen, ever more disorderly, multitudinous, radiational broadcastings of all nonsimultaneously disintegrating star and galaxy systems of eternally regenerative Universe remotely and nonsimultaneously intermingle their differently accelerated and differently aimed cosmic offcastings to produce nonsimultaneous entropic maelstroms variously distanced from one another: cosmic clouds.

In passing one another at a wide range of high velocities, these separate novae-refuse entities intermingle in a wide variety of densities. As they do so, their patterns interweave, accidentally thickening because their relative interattractivenesses vary inversely, according to Newton's interattractiveness law. Their interattractivenesses are countered by their respective velocity momentums, resulting in progressive veerings of the celestial courses of the individual items of cosmic refuse.

The interattractions produce progressively higher orders of gain and ever-decreasing radii of individual orbiting.

Gradually entropy gives way here and there to syntropy, as the individual components from a myriad of different stars gravitationally integrate here and there as new individual cosmic clouds. Within these clouds the process of orbital course veering into ever-lesser radii continues, and the cloud thickenings continue to seemingly endless progression.

In the same integrated interattracting and momentum veering of their orbital courses, all the separate, individually thickening clouds progressively converge here and there as larger and more complex clouds. These individual larger clouds progressively thicken together. This process leads eventually in sufficient condensations here and there in Universe to produce asteroids and planets.

Now repeating, now amplifying, it follows that the omniconserved, complexedly and nonsimultaneously intertransforming energies of eternally regenerative Universe consist most simply of two prime patterns: one of energy convergently associative as a complex mix of disorderly, cosmically broadcast, individually and multiplicatively disintegrating, asymmetric components in the course of the systematically organized and symmetrically converged, nonintertouching interarrays transformed into a radiantly entropic star whose randomly broadcast entities progressively intermingle in an even more disorderly manner with entropically (disorderly) broadcast energetic entities of a myriad of other as yet entropically broadcasting stars as well as with the still entropically traveling broadcastings of now energy-spent, only "has-been" stars. The superdisorderly cosmic intermingling of the energetic rubbish of vast numbers of stars sometimes interapproach one another to such a degree of proximity that Newton's celestial-bodies interattraction law progressively decreases the distances between them and thereby increases their interattraction at a second-power rate of the arithmetical distances intervening; thus, they palpitatingly interpulse now this way, now that way, finally collecting dominantly in zillions of initial trendings toward symmetries and responses in antientropic (i.e., syntropic) forces of orderly convergence. Gravitationally embraced, they begin interapproaching one another to become progressively electromagnetically and gravitationally self-sorting in terms of size and angularity and, in relative proximity, attaining here and there closest proximities of symmetric interpositionings which, when numerically and geometrically and vector-balanced in sufficient degree, finally attain the state of matter designed thereby to agglomerate with other newly formed matter until a critical mass of matter is convened, at which moment that matter suddenly starts to transform again, becoming a radiant star.

What is important to undestand and to concern ourselves with anew is the phase when only as maximum disorderliness occurs and only as myriads of disorderly, interdisposed groups of mistily lesser or foggily greater densities of disorderliness in many locations of Universe, tentative precloud pulsations occur on the myriads of nonsimultaneous critical thresholds of the moments-of-imminence of cosmically localizing, thereafter to become progressively denser, yet only locally within their interclearing and convergence, owing to varying dominances of radiational and gravitational forces, and, at first and for long, progressively disorderly for aeons vast quantities of energetic Universe are at all times preoccupied in this still disorderly limbo and thus still unconceptualizable vacillating condition, and ergo large quantities of cosmic energy are undifferentiated and not yet accountably associable as a certifiable entity.

After aeons of subcomponents interacting in the void of darkness, gradually, on the face of the deep, gravitationally collecting clouds once again began to appear and their relentless trending eventually in new creation would render once more cosmos out of chaos.

Synergetics reveals much more about the way energy quanta become temporarily vectorially lost to cosmic account, yet are realistically recoverable by Universe through the octahedron as conservation and annihilation model.

This pre-Magellanic-cloud, prestardust, preanything, for-the-time-being-inevident, nonconceptual, unimaginable, only-potentially-unlost, and only-in-pure-principle-recoverable phenomenon physically demonstrates my transformation model and shows the octahedron in its annihilation and conservation modes.

The vastness of overlapping unaccountability is difficult for those unschooled in synergetics to comprehend—resorting to explanations involving inverted energies, black holes, and latent phases—within an ever vastly regenerative Universe, with its multinonsimultaneity and, only in overall eternity, regenerativity.

7 INTEGRITY

THE MANY AND SEEMINGLY UNRELATED TOPICS we have reviewed are intimately interrelevant and seem to be operating synergetically. Einstein's greatness evolved from his synergetic concern for all experimentally verified data regarding the Universe and its progressively beknownst to us, ever more complexedly and nonsimultaneous physical intertransformings—associatively as matter and disassociatively as radiation.

At the very moment humanity has arrived at that evolutionary point where we do have the option for everyone to "make it," I find it startling to discover that all the great governments, the five great religions, and most of big business would find it absolutely devastating to their continuance to have humanity become a physical, metabolic, economic success. All the political, religious, and moneymaking institutions' power is built upon those institutions' expertise in ministering to, and ameliorating, the suffering, want, pain, and fears resultant upon the misassumption of a fundamental inadequacy of life support on our planet and the consequent misfortune of the majority of humans.

Some religious bodies battle politically and morally against abortion, which inherently eliminates their most lucrative raison d'être—humans and more humans and their concomitant adversity and suffering, and their need of ministry.

The institutionalized catering to want and suffering gives us a sense of the almost certainly fatal dilemma we are in. Another relevant threat to human continuance in Universe is our world education systems' deliberate cultivation of specialization, despite the fact that each individual human is born physically and metaphysically equipped to function as a natural comprehensivist, with a unique mind designed to ascertain and comprehend the generalized design principles governing interrelationships. Surely if nature had wished humans to be specialists, she would have given them the special integral equipment for so doing, as she has given to all other creatures.

How did it come about that the educational system was organized to counter this innate proclivity, environmental versatility, and multifaceted capability?

We have observed for aeons herds of wild horses led by a king stallion. Every once in a while an unusually big and powerful young stallion is born—much bigger than the other young stallions. When the big young one matures, the king stallion challenges him to a battle, with the winner inseminating the females of the herd. Darwin cited this as an example of nature's way of arranging to keep the strongest strains going.

I am sure that amongst the early human beings occupying our planet Earth, every once in a while a man was born much bigger than other men. He did not ask to be big; he just found himself to be born so. He found himself continually asked for favors. "Mister, I can't reach the bananas. Could you get some bananas for me?" Being good-natured, he would oblige. And then all the little humans around him said, "Mister, the people over there have lost all their bananas. They're dying of starvation. They're going to come over and kill us to get our bananas. You're big. You get out front and protect us."

So, the big man found his bigness being continually exploited. He said, "All right, people, you've got me out there fighting for you time after time, but between battles I'd like you to help me get ready for the next battle. I need weapons and walls."

The people said, "Okay, we'll make you king and you tell us what to do."

So, the big one became king. Another big stranger came along and said, "Mr. King, you have a soft job here. I'm going to take this away from you." The two battled. The king licked the stranger. The king had his opponent down on the ground and said, "You were going to kill me so you could have my kingdom, weren't you? You understand I can kill you right now, don't you? Okay, you're a very good fighter and I need a lot of good fighters around here, so, if you will promise to always fight for me, I'll release you.'' The stranger acquiesced. The king found himself to be an institution—a power structure.

The king then said spontaneously to himself, ''Don't let two big men come at me at once. I can handle them, but only one by one.''

From this instinct there gradually emerged a number-one grand strategy for all power structures: Divide to conquer. To keep conquered, keep divided.

The king said to himself, ''I want more of these big men. I'll make one the Duke of Hill A and the other the Duke of Hill B. Then I'll keep my spies watching to see that they don't gang up on me.''

Next, a whole lot of little people made trouble for the king by not obeying him. There were some very bright little people around. They refused to do what the king wanted done. The king had one or two of his big men bring in the little offenders. The king said, ''Mister, I'm going to cut your head off. You're a nuisance around here.'' The man replied, ''Mr. King, you're making a very great mistake cutting my head off.'' The king said, ''Why?'' ''Well, Mr. King, I understand the language of your enemy over the hill and you don't. I heard him say what he's going to do to you and when he's going to do it.''

The king said, ''Young man, you may have a good idea. You let me know every day what my enemy over the hill thinks he is going to do to me, and your head is going to stay on. Then you're going to do something you never did ever before—you're going to eat regularly, right here in the castle. We're going to put purple and gold on you so I can keep track of you.''

Then some other physically small character made trouble for the king, and it turned out that he could make better swords than anybody else. He was a great metallurgist. The king made him court armorer and had him live in the castle and wear purple and gold.

Somebody else made trouble and said, ''Mr. King, the reason I'm able to steal from you is because you don't understand arithmetic. Now, if I do the arithmetic here in the castle, people won't be able to steal from you.'' He, too, got the victuals, purple, and gold.

Speaking to all his little ''experts,'' the king said, ''You mind your business. And you mind your business. Is it clear to each and all of you that I'm the only one who minds everybody's business?''

The king now had all the great fighters, all the right intelligence, the right arms, the right logistics. His kingdom was getting very big. He wanted to leave it to his grandson. After years of success the king said to each of his experts, "You're getting pretty old. I want you to teach somebody about that mathematics. I want you to teach somebody about that metallurgy of yours," and so on.

Ultimately all the foregoing led to the founding of the educational-category scheming, as manifest in the organization of Oxford University and all other education institutions.

In spite of all humans' innate interest in the interrelatedness of all experience, long ago these world-power-structure builders learned to shunt all the bright intellectuals and the physically creative into specialist careers. The powerful reserved for themselves the far easier, because innate, comprehensive functioning. All one needs to do to discover how self-perpetuating is this disease of specialization is to witness the inter-departmental battling for educational funds and the concomitant jealous guarding of the various specializations assigned to a department's salaried experts on each subject in any university.

In the early 1950s, attending the American Association for the Advancement of Science annual congress in Philadelphia, I happened to find two papers that were presented in different parts of the symposium. One was in anthropology and the other was in biology. A team of anthropologists had for a number of years been examining all the known case histories of human tribes that have become extinct, and a team of biologists had been examining all the known cases of biological species that had become extinct. Both of the papers determined extinction to be the consequence of overspecialization.

How might this be? We know that we can inbreed ever-faster-running horses by mating two very fast-running horses—the mathematical probability of concentrating the fast-running genes is high. When you inbreed special ability, however, you outbreed general adaptability.

Its total energy being fixed and nonamplifiable, physical Universe uses that energy only rarely to do very big things—hurricanes, for example. Nature does smaller tasks more frequently and very small tasks very frequently. As human masters of highly bred racehorses inbreed the high-frequency everyday performance characteristics, they outbreed the rarely used survival capabilities. When the rare big-energy event occurs, the species, having lost its general adaptability to cope with unusual environmental conditions, often perishes. Quantum mechanics is the operating principle.

The energy of Universe may be divided into a few, very infrequent major events or into many, very frequent minor events. The energy of Universe is finite, and since multiplication is accomplished only by division, we recognize that quantum mechanics is the operating principle. Therefore, when humanity today is presented with the option of across-the-board success, it is so specialized as to be unable to recognize this generalized, only comprehensively discoverable and comprehendible course of action to be implemented by an invisible technology.

Humanity was given an enormous range of resources with which to discover that our minds are everything and our muscles are relatively nothing. We note that hydrogen does not have to "earn a living" before behaving like a hydrogen atom. Humans, in fact, are the only phenomenon upon which the power structure has been able to impose the everyday obligation of satisfactorily "earning a living."

Because of high technology's capability to take care of the needs of everybody on the planet, we now know that the prerequisite of having to earn a living is obsolete. Only by virtue of invisible technology's implementation of a revolution in producing constantly greater performance per unit of invested-resource-accomplished tasks has it come about that there are now adequate resources to take care of, and sustain, everybody at a high standard of living. Such a realization will swiftly alter the fundamental assumptions and activities of our daily lives in a very great way.

A preponderance of fear has long operated in the academic world amongst professional educators working toward, or holding tenuously onto, tenure. A great many teachers would gladly become research professors. If they were assured by some authority that they would be given the income they want, they would prefer to do much of their research and writing at home. Such home-conducted research and telecommuting among academics and other workers would save immense quantities of the irreplaceable fossil-fuel gasoline now used to commute daily to the workplace—especially in the United States.[1]

We must realize that we have all reached a turning point where we can no longer afford to make money rather than good sense.

Every child has an enormous drive to demonstrate competence. If humans are not required to earn a living to be provided survival needs, many are going to want very much to be productive, but not at those tasks they did not choose to do but were forced to accept in order to earn money. Instead, humans will spontaneously take upon themselves those tasks that world society really needs to have done.

If humanity realizes its potential in time to exercise this vital option, we shall witness strong competition among individuals to be allowed to

[1] A petroleum geologist friend of mine, François deChadenedes, once calculated that each gallon of gasoline produced by nature would cost $1 million to produce at the time and energy rates currently charged by utility companies.

serve on humanity's research, development, and production teams. Never again will what one does creatively, productively, and unselfishly be equated with earning a living. People's sense of accomplishment will derive from showing their peers and demonstrating to the great intellectual integrity of Universe, which we speak of as God, that they vastly enjoy doing their best in the unselfish production of service for others rather than just for the survival needs of themselves and their families.

I think all humanity has crossed the threshold to enter upon its "final examination." It is not the political systems or the economic systems but the human individuals themselves who are in final examination.

How much courage and integrity does each of us have individually to steer a life course according to what our minds have learned through experimental evidence to be the relevant principles governing our situation? How much ingenuity do we have to solve the larger problems of society through anticipatory design rather than through outmoded institutions based on misinformation and the maintenance of the status quo for the vested interests?

I have discovered that we have just such an option. How much courage does each of us have to take the first active step leading to the exercise of that option? What is it that each of us must do? How much willpower must we gather to cast aside deeply ingrained patterns and prejudices? How far must we go to make consciously considerate decisions based on intellectual integrity? How much faith must we have in our ability to recognize that intellectual integrity? Or, by default, will the unconscious crowd-following mass psychology of the Dark Ages reign supreme for another aeon?

When nature has an all-important function to be performed by any of her bioinventions and the chances of that biological invention surviving are poor, nature invents many alternative circuits to provide the same results. Nature is not depending solely on the intellectual courage and integrity of this one relatively minor team of human minds on our planet Earth to perform all the local Universe's information gathering and local problem solving. The intellectual integrity of Universe has myriads of alternate fail-safe ways of carrying on. Nature never puts all her eggs in one basket. This gives me reason to surmise that this particular Earthian team is in final examination and that its track record has been far from exemplary.

The human condition today has much improved from when I was young. Illiteracy was then overwhelming. The Soviet Union after the 1917 revolution, for instance, was 95 percent illiterate. For industrialization to work, that condition needed to be reversed, and it was.

The people I worked with on my first pre–World War I jobs were expert craftsmen and very kind human beings, but their on-the-job vocabularies consisted of no more than one hundred English words, almost half of them blasphemous or obscene. Today, the average six-year-old American child has a vocabulary of five thousand words.

The whole communication and information environment of humanity has undergone a revolution. Everybody world-round has a workable vocabulary today. This communications change has taken place at an incredibly rapid rate. In the last twenty-five years I have been around the world forty-eight times, and I am able to communicate wherever I go.

Nature has brought us to the communicating capability where we have 74,000 words in the *Concise Oxford Dictionary* and 150,000 words in most American "college" dictionaries. This proves that we have need of descriptive words for a great many unique experiences. That we could agree on the meaning of 150,000 words is extraordinary. We have reached the point where we are now possessed of sufficient information for each individual human to dare to exercise the option to "make it" rather than having to depend on the decisions of an educated elite.

In astrophysics we can access an omnidirectional 11.5-billion-light-year-radius reach for information. We have photographed the atom. We are at an evolutionary point where we should break out of our Dark Ages eggshell to act in a completely new and unexpected kind of way. A new emergent worldview provides us with clues about our wonderful new metaphysical environment.

Evolution may be classed into two types—what I call "number-one evolution" and "number-two evolution." Number-two evolution is operative wherever and whenever human beings think they are running the world. Number-one evolution is that in which nature is entirely responsible for the evoluting. Number-one evolution suggested in my lifetime that fallout from the comprehensive employment of the doing-ever-more-with-ever-less-resources-per-function invisible revolution by the military was entirely and unwittingly responsible for the fact that since 1900 we have gone from less than 1 percent to more than 65 percent of humanity enjoying a higher standard of living than had been ever experienced by any potentate when I was young. During that time we have also doubled the population, so we have actually increased by a factor of 130 the number of those so benefiting from this inadvertent technological fallout.

This fallout from political-economics doing the right things for the wrong reasons is what I mean by number-one evolution. There was no planning by any nations or enterprises that sought to alter the lives of all humanity in this historically unprecedented manner. Einstein loved humanity, but was dismayed at the lack of efficient planning to make everyone a success. Official planning was highly biased. There was no organized effort to improve the standard of living across the board. As we discovered earlier, it was assumed you could not, or must not, do so.

In 1938, I predicted in *Nine Chains to the Moon* that by the year 2000 the fundamental needs of everybody could be taken care of. I think Earthlings' final examination has been incrementally advanced and that there may not be that much time. The time remaining to switch over to a winning life strategy is less than a decade, possibly as short as three years. Every day and in every way humanity feels this crisis deeply. Talk of nuclear disarmament and dealing with environmental and social catastrophe is in the air.

Historically, females carrying the young in the womb could not cover as much geography as could the males. Females tended then to stay around a hearth, where they kept a fire going to cook the meals while the males hunted. Because he covered more territory and could report what he saw, man was also the news bearer. Dad could tell his children what he saw from the top of the nearby mountain. He could tell his children what the chieftain over the mountain was saying or doing.

All through history children, starting naked, helpless, and ignorant, have had Dad and Mom telling them what they could eat and what would and would not poison them. Parents told their children what they could and could not get away with in the power system. Dad and Mom were the authorities on how to get on. But Dad was also the authority who brought home the news. Dad's language was local and somewhat esoteric. The kids immediately emulated the way Dad spoke. He was the communication authority. New dialect after dialect was spawned.

Suddenly thrust into my world at age three was the invisible electron. No one took notice. When I was twenty-three, by virtue of that electron, we heard the first human voice on the radio. When I was twenty-seven, the first broadcasting station was licensed. In 1927, when I was thirty-two, all the dads around the country came home one evening to the kids' excited imperative, "Hurry, Daddy, listen to the radio! A man is trying to fly across the Atlantic." Dad said "What!" and never again was the one to bring home the news.

Nobody thought about this event as a *number-one* anthropological evolution event. The kids knew that Dad and Mom were their private-home authority all right, but quite clearly, Dad and Mom ran across the hallway and got the neighbors to tune in the radio because the radio was going to tell them something important. The children observed for themselves that the radio was more of an authority than was either Dad or Mom. These greater authorities—the radio people—got their jobs on the radio by virtue of the commonality of their diction rather than the esoteric way that Dad said things. The radio people also got their jobs by virtue of the size of their vocabulary and versatility in employing it. To hold their jobs, they had to make their programming ever more popularly understandable, so they developed ever more precise vocabularies and ever-clearer enunciation.

As Dad and Mom accepted the radio-amplified authority, the kids emulated the speech styles of the people on the air. Noting this, many parents also adopted the radio speech, not wishing to be belittled in their children's estimation. This is what overnight changed the speech pattern of humanity the world around, even in the tiniest of hamlets. This was number-one evolution—not planned by humans—but altering human interrelations nonetheless.

The speed of sound is approximately 700 miles per hour, given an average temperature. The speed of electromagnetic radiation is 700 *million* miles an hour. Sound waves go no farther than the atmosphere. Radiation goes on and on (without atmosphere) in the Universe, giving us the infinite television views of distant planets remotely transmitted by solar-system-traversing satellites. The amount of information we can get with our eyes is a millionfold greater than what we can get with our ears.

In the mid-1960s, students at the University of California at Berkeley made the world news as the first dissidents in the university educational system. The Berkeley students asked to meet with me. That same year, I was also asked to speak to many of their contemporaries at other universities. In the last half century I have been invited to speak at over 550 universities and colleges around the world. At Berkeley I discovered that the 1965 dissidents were born the year television came into the American home.

The students said, "Dad and Mom love me to pieces. I love them to pieces, but they don't know what's going on." That was exactly the opposite of the way things were when I was young.

My father died when I was very young. My mother said very often, "Darling, never mind what you think. Listen, we're trying to teach you." At school they said, "Never mind what you think. Listen, we're trying to teach you." It was the assumption on the part of the pre–World War I older people that young people's thinking was utterly unreliable.

In 1965, I was fascinated to hear the young world suddenly saying, "Dad and Mom come home from the shoe store and have a beer. Then they watch television, but they have little interest or connection with humans going to Korea or Vietnam or to the Moon. They obviously don't have anything to do with anything. We can see that the people around the world are in great turmoil. We are going to have to do something effective in eliminating that trouble, since Dad and Mom have no understanding of, or concern with, the world's problems.''

That 1965 young world's compassion was suddenly of worldwide scope. It could never again be reduced to concern with only themselves and local issues.

The young world was saying, ''Dad and Mom don't understand what's going on, so I've got to do my own thinking.'' Nobody said to them anymore, ''Never mind what you think.'' They begin to think—earnestly, cautiously, individually—and then to test that thinking collectively. Because they did so, they became highly idealistic. They had no experience at taking the thinking-initiative, so they necessarily made mistakes.

The Soviet Union and the United States today spend over $400 billion a year to ready themselves for war. Of that amount approximately $20 billion a year goes for psychoguerrilla warfare—how to break down each other's (and third-party countries') economy and morale before arriving at the point of war by distribution of narcotics, social engineering, political movements, electronically amplified brainwashing, wheeling and dealing of various sorts. Young people's spontaneous thinking is idealistic. In the 1960s, that idealism was sometimes exploited. Quite often the gentle, angry young people discovered that their heads were sometimes used as battering rams rather than for thinking. Then, over the next fifteen years, through experience, they gradually matured, developing an immunity to political and social exploitation.

As I see it now, every child is born successively in the presence of a little less misinformation and in the presence of a great deal more reliable information. The young world is enormously advantaged.

I asked a young man in Pennsylvania who had written an extraordinary book on the Three Mile Island incident to visit me at my Philadelphia office. He was a high school dropout. I said, ''How did you get to writing?'' I have never read anything more interesting and sustaining than his book on Three Mile Island. It was well informed on all the bureaucratic decisions in Washington, all the mechanisms of the power structure. It was incredibly well done. He said, ''Well, I love reading. I liked particularly Shakespeare, Walt Whitman, and Mark Twain.'' He had quite a range of inspirers—he just loved them—and he had learned how to express himself well. He also had learned how to put relevant information together. He was typical of a young world that is progressing relentlessly. At twenty-one, he was neither misinformed nor misled. I was astonished. He seemed to me to be a heartening manifest of number-one evolution.

Each year, I get letters from children born after humans landed on the Moon. How these young ones find me to be somebody to write to, I do not know, but they do. They say that they understand that I may empathize with *their* concern. The letters are written in superb English. They are familiar with all the tasks that were necessary to get humans to the Moon and back safely. They are familiar with the Apollo Project's critical path. They know that humanity can do anything it needs to do. They wonder, "Why can't we set about to make this planet Earth work?" The young world gives increasing evidence of this level of concern. The after-the-Moon-landing young people will, within a few years, be able to take over the course-setting tasks of humanity as local Universe information gatherers and local Universe problem solvers in support of the integrity of an eternally regenerative Universe.

In 1979 a newspaperman in Los Angeles, Richard Brenneman, arranged for me to meet with a group of very young people to discuss the subjects I have dealt with in this book.

After six months of reading my books, each had prepared a set of questions about my thoughts and statements. They had lively interest in what I had to say. I asked them their ages. The oldest, a boy, was twelve. He said he was interested in learning the tricks of magicians. The next-oldest, an eleven-year-old boy, said that he was interested in electromagnetics. The third member of the group was a little girl who was then ten years old and the only one of these three born after humans had reached the Moon. I asked her in what she was interested. She answered, "I am a comprehensivist, like you. I am interested in everything." All youth born since the 1969 Moon landing are deeply familiar with the appropriation of billions of dollars for the complex technology of the Apollo Project. The Moon that for three million years had represented the unreachable had been successfully reached. The post-Moon-landers say, "Humanity can do anything it sets out to do. We need to make the world work for everybody on the planet. Let's get going." When they find out that I have discovered what can be technologically accomplished, they perk up their ears and roll up their sleeves.

The passion to understand engenders the passion to demonstrate competence. This is about to be demonstrated by that emergent young world.

An unprecedented transformation of all of our affairs is on the horizon. We are about to see all of the more than 150 nations of the world almost imperceptibly vanish, their function outmoded, their selfish and short-term pursuits no longer welcome or workable in an increasingly interdependent world economy. These nations represent more than 150 blood clots impeding the free circulation of the world's metals and the technological advantaging that they implement. When we engage the economies of recirculating all the metals as scrap, the entrenched mining interests will no longer be able to block that free flow. With the vast uncensorable network of communication media, obstacles to the free flow of vital information will become progressively more difficult to erect and enforce. Traditional human power structures and their reign of darkness are about to be rendered obsolete.

Revolutionary changes in every sphere of life must happen, and there is a young world very glad to realize them. I see clearly that intellectual integrity is trying to make humanity a success.

When I first began doing my own thinking in 1927, I said that I was going thenceforward to completely and irretrievably abandon everything I had ever been taught to believe—and, from that time forward, base my decisions only upon my own experimental evidence. It should be the vow of every scientist.

It is a prominent part of everyone's experience that enormous numbers of humanity are deeply moved by some religious credo or another. People manifest a deep sense that something is everywhere operative which is mysteriously greater than that which is negotiable by the knowledge and will of humans.

I constantly ask myself, "Do you have any experientially evidenced reason to assume a greater intellect to be operating in Universe than that of humans?" I answer myself, "The only-by-mind-discovered generalized principles of science that can only be expressed mathematically and mathematics are inherently intellectual." I found that I was overwhelmed by the experiential evidence of a cosmic intellectual integrity at work in the design of Universe. Thus, when I said in 1927 that I was going to try to find out and support what the great cosmic intellectual integrity was trying to do, I committed myself as completely as humans can to absolute faith in the wisdom of the eternal intellectual integrity we speak of as God. In 1930 Einstein's publication of his "Cosmic Religious Sense," which described his "nonanthropomorphic concept of God," told me that the most profoundly thinking human on our planet was also so committed.

At the outset of my 1927 commitment, I realized that my exploration for comprehension of God's design of eternally regenerative Universe might well mean that I could develop some very powerful insights. I asked myself if I could trust myself never to turn the power of such insights to personal advantage. Never to consider myself special vis-à-vis God. Never to develop a cult. Never to exploit for selfish reasons the insights I was sure to experience in operating an enterprise backed only by intellectual integrity. My answer to myself was, "Yes, I can trust myself not to selfishly exploit the power of cosmic insights." I have kept my promise, which brings me back to my opening statement about myself: I am an average, healthy human—no less, no more. But all average human beings are magnificently endowed with creativity, and mysteriously capable of vastly more than any of us has ever assumed to be possible.

While it is possible to recognize that humanity is still comprehensively locked in by the Earthian power structures' Dark Ages conspiracy, it is as yet not possible to assess exactly how powerful that imprisonment is. The fact that a vast number of humans still assume that it is within the power of their political leadership and the military might they command to resolve our problems is a reasonable manifest of the continued imprisonment of all humanity.

To this author, the dilemma is so great that in 1983 he found himself writing the following paragraphs, which he titled "Integrity":

A VERY LARGE NUMBER OF EARTHIANS, possibly the majority, sense the increasing imminence of total extinction of humanity by the more than 50,000 poised-for-delivery atomic bombs. Apparently no one of the 4.5 billion humans on our planet knows what to do about it, including the world's most powerful political leaders.

Humans did not invent atoms. Humans discovered atoms, together with some of the mathematically incisive laws governing their behavior.

In 1928, humans first discovered the existence of a galaxy other than our own Milky Way. Since then we have discovered 100 billion more galaxies, each averaging over 100 billion stars. Each star is an all-out chain-reacting atomic energy plant.

Humans did not invent the gravity cohering the macrocosm and microcosm of eternally regenerative Universe.

Humans did not invent humans or the boiling and freezing points of water. Humans are 60 percent water.

Humans did not invent the ninety-two regenerative chemical elements or the planet Earth with its unique biological life-supporting and protecting conditions.

Humans did not invent the radiation received from our atomic energy generator, the Sun, around which we designedly orbit at a distance of 93 million miles.

The farther away from its source, the less intense the radiation. With all the space of Universe to work with, nature found 93 million miles to be the minimum safe remoteness of biological protoplasm from atomic radiation generators.

Humans did not invent the vast, distance-spanning photosynthetic process by which the vegetation on our planet can transceive the radiation from the 93-million-mile-away Sun and transform it into the complex hydrocarbon molecules structuring and nurturing all life on planet Earth.

Design is both subjective and objective, an exclusively intellectual, mathematical conceptioning of the orderliness of interrelationships.

Since all the cosmic-scale inventing and designing is accomplishable only by intellect, and since it is not by the intellect of humans, it is obviously that of the eternal intellectual integrity we call God.

All living creatures, including humans, have always been designed to be born unclothed, utterly inexperienced, ergo absolutely ignorant. Driven by hunger, thirst, respiration, curiosity, and instincts such as the reproductive urge, all creatures are forced to take speculative initiatives or to "follow the herd," else they perish.

Ecological life is designed to learn only by trial and error.

Common to all creature experience is a cumulative inventory of only-by-trial-and-error-developed problem-solving reflexes.

Unique to human experience is the fact that problem solving leads not only to fresh pastures, but sometimes to ever more intellectually challenging problems. These challenges sometimes prove to be new, more comprehensively advantaging to humanity, mathematically generalizable, cosmic design concepts.

Humans have had to make trillions of mistakes to acquire the little we have thus for learned.

The greatest mistake we have ever made is to assume that the supreme authority governing life and Universe is not God but either luck or the dicta of the humanly constituted and armed most-powerful socioeconomic systems and religions. The combined human power structures—economic, religious, and political—have compounded this primary error by ruling that no one should make mistakes and punishing those who do. This deprives humans of their only-by-trial-and-error method of learning.

The power structure's forbiddance of error-making has fostered cover-ups, self-deceit, egotism, false fronts, hypocrisy, legally enacted or decreed subterfuge, ethical codes, and the economic rewarding of selfishness.

Selfishness has in turn fostered both individual and national bluffing and vastness of armaments. Thus, we have come to the greatest of problems ever to confront humanity: What can the little individual human do about the supranational corporate power structures and their seemingly ungovernable capability to corrupt?

A successful U.S. presidency campaign requires a minimum of $50 million, senatorships $20 million, representatives $2 million. Through big business's advertising-placement control of the most powerful media, money can buy, and has now bought, control of the U.S. political system once designed for democracy.

Without God, the little individual human can do nothing. Brains of all creatures, including humans, are always and only preoccupied in coordinating the information fed into the brain's imagination—image-I-nation—its scenarioing center, by the physical senses and the brain-remembered previous similar experience patterns and the previous reflexive responses.

Human mind alone has been given access to some of the eternal laws governing physical and metaphysical Universe, such as the laws of leverage, mechanical advantage, mathematics, chemistry, and electricity, and the laws governing gravitational or magnetic interattractiveness, as manifest by the progressive terminal acceleration of Earthward-traveling bodies or by the final "snap" together of two interapproaching magnets.

Employing those principles first in weaponry and subsequently in livingry, humans have been able to illumine the nights with electricity and to intercommunicate with telephones and to integrate the daily lives of the remotest-from-one-another humans with the airplane.

As a consequence of human mind's solving problems with technology, within only the last three-fourths of a century of our multimillions of years' presence on planet Earth, the technical design initiatives have succeeded in advancing the standard of living of the majority of humanity to a level unknown or undreamed of by any pre-twentieth-century potentates.

Within only the last century, humanity has grown from 95 percent illiterate to 65 percent literate. Preponderately literate humanity is capable of self-instruction and self-determination in major degree.

Clearly, humanity is being evolutionarily ejected ever more swiftly from all the yesteryears' group-womb of designedly permitted ignorance.

Regarding the power-structure-supported Scriptures' legend of woman emanating from a man's rib, there is no sustaining experiential evidence. Humanity now knows that only women can conceive, gestate, and bear both male and female humans. Women are the continuum of human life. Like the tension of gravity-cohering, space-islanded galaxies, stars, planets, and atoms, women are continuous. Men are discontinuous space islands. Men, born forth only from the wombs of women, have the function of activating women's reproductivity.

The present evolutionary crisis of humans on planet Earth is that of a final examination for their continuance in Universe. It is not an examination of political, economic, or religious systems, but of the integrity of each and all individual humans' responsible thinking and unselfish response to the acceleration in evolution's ever more unprecedented events.

These evolutionary events are the disconnective events attendant upon the historic termination of all nations. We now have 163 national economic "blood clots" in our planetary production and distribution systems. What is going on is the swift integration in a myriad of ways of all humanity not into a "united nations" but into a united space-planet people.

Always and only employing all the planet's physical and metaphysical resources only for all the people, this evolutionary trend of events will result in an almost immediately higher standard of living for all than has ever been experienced by anyone.

In general, the higher the standard of living, the lower the birthrate. The population-stabilizing higher living standards will be accomplished through conversion of all the high technology now employed in weaponry production being redirected into livingry production, blocked only by political party traditions and individually uncoped-with, obsoletely conditioned reflexes.

A few instances of persistent, misinformedly conditioned reflexes are the failure popularly to recognize the now scientifically proven fact that there are no different races or classes of humans; the failure to recognize technological obsolescence of the world-around politically assumed Malthus-Darwin assumption of an inherent inadequacy of life support, ergo "survival only of the fittest"; the failure to ratify ERA, the equal rights (for women) amendment, by the thus-far-in-history most-crossbred-world-peoples' democracy in the U.S.A.; or, with ample food production for all Earthians, the tolerating of marketing systems which result in millions of humans dying of starvation each year.

Carelessly unchallenged persistence of a myriad of such misinformed brain reflexings of the masses will signal such lack of people's integrity as to call for the disqualification of humanity and its elimination by atomic holocaust.

You may feel helpless about stopping the bomb.

To you, the connection between the equal rights amendment and the atomic holocaust may at first seem remote. I am confident that what I am saying is true. The holocaust can be prevented only by individual humans demonstrating uncompromising integrity in all matters, thus qualifying us for continuance in the semidivine designing initiative bestowed upon us in the gift of our mind.

THE BEST ANTIDOTE to the powerfully misintentioned sensing and acting reflexes of society is the study of synergetics. The data of synergetics as presented in the two volumes of *Synergetics* and background data in *Critical Path* (1981) are adequate to the task of breaking the Dark Ages stranglehold on the human individual. This book has undertaken to present some of the principal synergetics concepts in a logical sequence. It does not treat the successively acquired realizations in the detailed degree of *Synergetics,* the definitive reference on the subject.

As a guidebook to synergetics in the context of its historic roots, this volume has added new insights and primary concepts to the subject, including some that have accrued since its earlier expositions. Study of synergetics with continued recommitment of human individuals to utter faith in the comprehensive wisdom and absolute power of the intellectual integrity and love governing an eternally regenerative Universe may bring about our ultimate escape from the Dark Ages' race-suicidal obsession with the misconception that cosmic supremacy is vested in little planet Earth's politicians, priests, generals, and monetary-power wielders.

Dear reader, traditional human power structures and their reign of darkness are about to be rendered obsolete.

For further information on Buckminster Fuller and synergetics:

BUCKMINSTER FULLER INSTITUTE, 1743 S. La Cienega Blvd., Los Angeles, CA 90035. Maintains the Fuller archives and "Chronofile," presents educational programs, publishes *Trimtab* quarterly newsletter. Write for mail order catalog, listing books, educational materials, maps, and other items. Phone: (310) 837-7710, FAX (310) 837-7715.

CRITICAL PATH PROJECT, 2062 Lombard St., Philadelphia, PA 19146. Information exchange, research on applications of Fuller's geometry, operates conferencing computer bulletin board system for exchange of ideas on synergetics and study of Fuller-related topics. Write Kiyoshi Kuromiya for further information. Phone: (215) 545-2212, FAX (215) 735-2762, computer BBS (215) 564-1052 (4-lines).

SYNERGETICS INSTITUTE, 680-345 Tomo, Numata-cho, Asaminami-ku, Hiroshima, Japan 731-31. Research and educational program on synergetics, produces Hypercardstackware of 3-D animations of the process of making hierarchies of the icosahedron and the rhombic triacontahedron. Write Yasushi Kajikawa for further information. Phone: (082) 848-3539.

INDEX

Note: Numbers in *italics* refer to figures and their captions.

Abacus, 34
Abortion, 248
Acceleration. *See also* Angular acceleration
  Galileo's experiment on, 82, 84
  into orbit, 86
Air, identification of gases in, 34–35
Air turbines, 23, *24–26*
Alexander the Great, 95, 107
Alexandria (Egypt), 95, 96, 100, 102
Algebra, 118
Alloying, 110–11, 113, 146, 238
  increased tensile strength through, 155
  quasicube in, 156–57, *156, 157*
Allspace
  coordinating systems for, 52–53, 58, 60
  quarks as minimum fillers of, 232
  rhombic dodecahedra as filling, 59, 170–71

Aluminum, 109–10
American Association for the Advancement of Science, 251
American Geographical Society, 235
American Institute of Architects (AIA), 14
  ideal 1927 house of, 112–13
Angles, 171–73, *173*
  sum of, around system, 191–93
Angular acceleration, 19–27, *22, 24–27,* 209–13, *212*
Antientropy, 74. *See also* Syntropy
Apollo Project. *See* Moon—astronauts' landing on
Arabia, ancient, 94
Arabic numerals, 34, 102
Archimedes, 39
Aristarchus, 100
Around, 104
Artifact revolution, 7–9
  advantages to humanity from, 3, 14–15
Artifacts. *See also* Inventions
  definition of, 8
  problem-solving by, 8–9

Assumptions, elimination of, 30
Asteroids, Universe's creation of, 245
Atmospheric reentry heat, 174
Atomic bombs
    dilemma of, 260
    disarmament of, 255, 264
Atomic clock, 49
Atomic energy, phasing out of, 116,
    218
Atoms, 39. *See also* Nucleus; Quarks
    (mites)
    distance of electrons from nuclei in,
    130
    first conception of, 35
    in four-ball tetravertexion systems,
    148, *149*
    pulsing of, 49, 148
Atom smashers, 244–45
Austronesia, 91, 93, 98
Avogadro's law, 40, 118, 163, 232
Axial rotation, 18
Axiomatic, definition of, 119
Axioms, elimination of, 30

Babylonia, 92–94
Balloons, 174–75
Bell-jar experiment, 34–35
Berkeley, University of California at,
    256
Betweenness, 131, *134*, 135
Big bang theorists, 38, 40
Bio-"life," 241
Black Mountain College, 88
Boeing 747 example, 6–7
Boggs, Samuel W., 235
Bowditch curves. *See* Lissajous figures
Boyle's law. *See* Avogadro's law
Brain, mind distinguished from, 4, 32–
    33, 36–38, 71, 84, 237, 262
Brenneman, Richard, 258
Bridgman, Percy Williams, 30
British Empire, 84–85
    as first spherical-world empire, 107
Brownian movement, 80, 83, 106
Buckminsterfullerenes, 67*n*
Buddhism, 95–96
Butlerev, Alexander, 220

Calculus, 197
    Newton and, 84
California, University of, at Berkeley,
    256
Canvas, wood-frame-mounted, 143, *143*

Capitalism, 108–9
Carbon, 67*n*, 110, 184, 220, 221
Celestial bodies. *See also* Planets; Stars
    human observation of, 35–36
    physical laws of, 4–5, 20, 33, 36–37
c-g-s system, 220, 230
Christianity, 96–97. *See also* Roman
    Catholicism
Chrome-nickel-steel alloy, 155
Cipher, invention of, 34, 35, 37,
    102–3, 118
Circles. *See also* Great circles
    as finite polygons, 192
Clipper ships, 13–14
Club of Rome report (1974), 115
Columbus, Christopher, 103
Columns
    Greek, diameter-to-length ratio of,
    240
    possible height of, 88
Communism, 108–9
Compasses, ships', 213
Compression and tension, 87–88, 109,
    234
Computers, trigonometric, 197
Concave and convex, 48, 51, 57–59,
    137–38, *137*, *138*
Considered, consideration, definition
    of, 71*n*, 130
Coordination systems
    cubical, 52–53, 60
    tetrahedron-based, 52–55, 58
    XYZ, 63, 99, 140, 196, 197, 220,
    228
Copernicus, Nicolaus, 33, 35, 103
Copper, 114, 115
    hemispheres spun in, 236
Corners (vertexes), 141, *142. See also*
    Euler's formula; Flex-corners
Cosmic clouds, 245–46
Coupler, *61*
    definition of, 58
    description and function of, 61–62
Crates (philosopher), 100
Critical path, 115
Critical Path (book), 264
Critical proximity, 86–87, *120*, 208
Critical speed, 86–87
Crystallography
    basic units of, 55, 67
    triple-bonding in, 163, *168*, 169, 239
Cube, 52–53, 60, 129, 138
    angular topology of, *198*

in conventional geometry, 223–24, 230
new name for, *132*
as nonstructural, 146, *147, 157–58, 158–60,* 221
stabilized by tetrahedron, 222, *223*
in synergetic mathematics, 224
in synergetics' hierarchy, *128–29*
tensegrity, *179*
tensegrity stacked, *185*
tetrahedral mensuration applied to, *199*
of water, as unit of volumetric measurement, 166
Cubing, "tetrahedroning" preferred to, 226, *227*
Cubo-octahedron, *128–29*
Curl, Robert, 67*n*

Dark Ages, 1–2, 63, 66, 78, 82, 84, 85, 102–6, 118, 253, 260, 264
influence of, on modern "legal and academic" precedents, 111
Darwin, Charles, 108, 249, 263
deChadenedes, François, 252*n*
Democracy, 109, 262, 263
Democritus, 35, 130, 221
Design, meaning of word, 40
Design science, 240
anticipatory, 17
function of, 8
Diamonds, *145*
synthetic, 220–21
Digital calculators, 34
Dillard, Annie, 121
Divergent openings, *142*
Division by multiplication, 138
Dodecahedron, 138
angular topology of, *198*
pentagonal, *54*
new name for, *132*
tetrahedral mensuration applied to, *199*
rhombic, 59–61, *59,* 170–71
new name for, *132*
in synergetics' hierarchy, *128–29*
in synergetic mathematics, 224
tetrahedral mensuration applied to, *199*
Dorians, 93
Dreams, 57
Dymaxion constants, 223

Dymaxion House, 113
Dymaxion World Map, 93, 235–36

$E = mc^2$, 40, 76, 78, 79, 81, 118, 226
Earth. *See also* Ecology
as center of Universe, 98, 100–101, 103
as "entropy violation," 75
great circles of, 192
orbit of, *208–9*
Ecology
definition of, 9
of Earth as subset of that of Universe, 9–10
interdependent nature of, 15, 16
syntropic functioning of, 74–75
Education
holistic considerations broken up by, 37
specialization cultivated by, 249, 251
Egypt, ancient, 91, 93–95, 98
Einstein, Albert, 2, 29–30, 37, 38, 40, 75–87, 91, 111, 118, 220, 241, 248
atomic bomb and, 79
author's talk with, 78–79
"The Cosmic Religious Sense: The Non-Anthropomorphic Concept of God," 75–76, 259
experimental evidence for theory of, 77–78
interest in tensegrity sphere by, 87, 88
Newton's approach compared to that of, 38, 80–86, 106
scenario Universe discovered by, 105–6
Electric power transmission, 218
Electromagnetic waves
conceptual picture of, 57
discovery of, 35
as not continuous, 232
in solid-state physics, 66
Energy
atomic, phasing out of, 116, 218
cosmic, speed of light as norm for, 80
electrical
always follows highest tension, 234
proposed worldwide grid of, 218–19

Energy (*cont.*)
fossil fuels to end as source of, 112, 116, 218
great circles for transmission of, 67, 232–33
in iceberg-melting, 238
loss and gain of. *See* Entropy; Syntropy
Newton's conception of, 80–82
relative size and effectiveness of, 239
required by cube vs. tetrahedron, 52–53
solar, 116
in subatomic particles, 61–63
of Universe, as finite, 251
Entropy, 51, 73
cosmic clouds and, 245–46
in Newton's physics, 81
quanta modules and, 56–57
Environment. *See also* Ecology
development of more favorable, 3
reforming of, 8
Equal rights amendment (ERA), 263–64
Eratosthenes, 100
Eternity, 5, 38, 50, 64, 169
Euclidean geometry, 97–101
Eudoxus, 100
Euler, Leonhard, 39, 118, 131, 143, 213, 217–18
Euler's formula (law), 58*n*, 122, 127, 138–39, 144, 161
restating of, 134–35
Events
lines as trajectories of, 233
and multiplication by division, 237–38
in occupation of points, 232
Evolution
human, new stage of, 15–16, 18, 28–29, 254–55, 262–63
involution and, 208–9, *210, 211*
two types of, 254
Exceptionlessness, 5, 38, 64
Expansion-contraction, 18
Experiences
as components of Universe, 127
generalized or special-case, 70–71
omnirecallability of, 64
as systems, 131
Experiments
decisions should be based on, 259
environmental conditions of, 30

Extinction, as consequence of overspecialization, 251–52, 260
Eyeglasses, principle of, 12, 65–66

Faces, 141, *142. See also* Euler's formula
Family, recent change in nature of, 255–57
Fermi, Enrico, 78, 221
Flex-corners (flexible corners), 47, *48, 146, 147,* 225
"Fluxions," 192
Flywheels, 23, *26*
Fossil fuels, 112, 116, 218
star-igniting, 73
Four-dimensionality, 50, 144, 205, 208, 230
Four-vertexion, *132. See also* Tetrahedron
Frankland, Sir Edward, 220
Fullerenes, 67*n*

Galaxies, discovery of, 260
Galileo, 5, 13, 33, 37, 75, 82–86, 103, 118, 238
laws of motion of, 85
Gases
in air, 34–35
same number of molecules per given volume of, 232
as single-bonded, 163, *168,* 169, 239
in tensegrity models, 174–82, *181*
touching by, 232
Gautama Buddha, 95
General Electric Laboratories, 220–21
Generalizations, as metaphysical and eternal, 71
General System Theory, 219
Genghis Khan, 107
Genius, 37, 124
Geodesic domes, 14, 217, 236
number installed, 115
Geodesics
definition of, 135, 192
tetrahedral mensuration applied to, *199*
Geodesic tensegrity sphere, 87–90
Geometry
of atomic physics, 194
early navigation by, 91–93, 96–98
four-dimensional, 50, 144, 205, 208
Greek, 97–100, 130, 193, 213, 217
physics divorced from, 46

synergic. *See* Synergetics
  traditional, 219, 223–25
  unconventional, 10–13
Gibbs's phase rule, 40, 118, 163, 166
Gimbal system, *212,* 213
Glove, inside-outing of, 205, *207*
God, 1–2, 59, 137
  emergence of belief in one, 95–96
  as eternal sum of all truths, 169
  as intellectual integrity of Universe,
    253, 259, 261
  non-anthropomorphic concept of, 75–
    76, 259
Goddard, Robert, 86
Government, origin of, 249–51
Grasshopper, jumping of, 239
Gravity
  coexistence of radiation and, 39, 51,
    56–57, 89–90, *90, 200*
  as coherence, 56
  as conserver of integrity in Universe,
    52
  discovery of, 5, 36–37
  linear acceleration of, 212
  Newton's law of, 5, 37, 145–46,
    150–51, *152*
Great circles
  definition of, 192, 233
  for energy transmission, 67
  foldable, 233–35, *234*
  as information-shunting and -holding
    circuits, 67–68
  in rotation of icosahedron, *203*
  secondary sets of, 68
  spherical, *196*
    as geodesics, 192
  spinning of, *54,* 55
  in tensegrity spheres, 177
  three-way grid of, 235–36
Greece, ancient, 93, 95, 187
  geometry in, 97–100, 130, 193, 213,
    217
Gyroscopes, 18, 19, *27,* 213

Hahn, Otto, 79
Hammer throwers, 19–20, 22–23, *22,*
  24, 209–12
Heracleides, 100
Hertz, Heinrich, 35
Hexahedron, 226
Hipparchus, 100
Houses, prefabricated, air-deliverable,
  113

Hubble, E. P., 6
Human affairs, abnormal trending of,
  from Newton's no-change norm,
  81–83
Human behavior, abandoning of reform
  of, 8–9
Human cells, tensegrity in architecture
  of, 185–87
Human mind, 4–6, 32–40, 71, 84, 116,
  237, 252, 262
  as syntropic, 73
Humans
  new stage of evolution of, 15–16,
    18, 28–29, 254–55, 262–63
  if not required to earn a living,
    252–53
  possible extinction of, 252, 260
  trial-and-error problem-solving by,
    261
  two prime motivations of, 29
  why included in design of Universe,
    2, 3–4, 6–7, 40, 116
Hummingbird energy requirements, 239
Hypercube, 50, 230

Ice, floating on water by, 232
Iceberg, melting of, 238
Icosahedron, 41, 50–51, *54,* 138, 174,
  228
  angular topology of, *198*
  axes of rotation of, *203*
  discovery of, 92
  great circles of, 235
  historical loss of concept of, 97
  as information shunter-holder, 67–68
  new name for, *132*
  spherical, 190, *197*
  in synergetics' hierarchy, *128–29*
  tensegrity, *175*
  tetrahedral mensuration applied to, *199*
In and *out,* 104
India, 91, 102
Indonesia, 98
Industrialization, literacy and, 253
Information revolution, 254, 257, 259
Ingber, Don, 185
Insideness-outsideness, 57–58, 70, 124,
  131, 169, 217
Inside-outing, 18, *43,* 49–50, *49,*
  205–8, *206, 207*
Interattractiveness of two bodies, Newton's law of, 5, 37, 146–48, 150–
  51, *152, 245, 246*

Interference phenomena, *120*
Interrelationships, number of, of any
given number of entities, 64,
132–33
Inventions
author's, 7, 14
gestation periods of, 7–8, 77
Invisible interrelationships, discovery
of, 33–39
Invisible technological reality, 109–11
Involution and evolution, 208–9, *210,
211*
Isotropic vector matrix. *See* Vector matrix, isotropic

Jeans, Sir James, 131
"Jitterbug," 11, 57, 67, 67*n*, 230, *231,*
240, 244

Kepler, Johannes, 4–5, 33, 36, 75, 82–
84, 86, 118, 241
"Killingry," 8*n,* 28
Kingship, in prehistory, 249–51
Kroto, Harry, 67*n*
al-Kwarizmi, 102

Lavoisier, Antoine-Laurent, 34–35
Laws of the Universe, 3, 5–6, 10, 29–
30, 33, 38–40, 64, 118
Leibnitz, Gottfried, 84, 192
Leverage, principle of, 46, 47
Life, 57, 59
beginning of, 70, 124, 135
Life support
in Malthusian theory, 107–9, 111, 263
war technology and, 17–18, 28, 29,
113, 115
Light
angularity in transmission of, 12
speed of, 38, 79–83, 85, 118
Light-years, 38, 105
Limits-to-growth theories, 108, 115
Linear acceleration, 19–22, *21,* 212
Lines
cannot pass through same point at
same time, 10–13, 119, *120*
geodesic, definition of, 192
of systems, 141, *142. See also*
Euler's formula
called "edges" or "vectors," *142*
as tetrahedra of great altitude and
almost zero base, 239–41, *242*
as trajectories of events, 233

Liquids, as double-bonded, 163, *168,*
169, 239
Lissajous figures (Bowditch curves),
232, 233, 235
Literacy, expansion of, 253–54, 262
"Livingry," definition of, 8*n*
Living standard
potential rise of, 7, 30–31, 111, 114,
115, 254–55, 263
recent rise in, 262
Love, 219
definition of, 39

MacCready, Paul, 45
Mach, Ernst, 131
Madagascar, 94
Malthus, Thomas, 107–9, 111–12, 115,
263
Manhattan Project, 79
Marx, Karl, 108
Massachusetts Institute of Technology,
115
Mathematics, 39
useless studies in, 118–19
Metaphysical, author's use of word, 48,
64–65
Metaphysical wonders, list of, 118
Michelson, Albert, 79, 80
Millikan, Robert A., 118
Mind. *See* Human mind
Mining, as now unnecessary, 115–16
Mites. *See* Quarks
Möbius, August, 143
Modulated noninterference, *120*
Mombasa, 94
Money, invention of, 17
Moon, 35, 37, 86, 101
astronauts' landing on, 104, 114–15,
258
Motion, Newton's first law of, 80–81
Motions, six fundamental, 199–205
Mouse falling from airplane, 13, 238
Multiplication by division, 51–52, 52*n,*
69, 138, *139,* 237–38
Murchie, Guy, 13, 238–39, 241

Nations, coming termination of, 259, 263
Navigation, early, 91–93, 96–98
Newton, Isaac, 30, 33, 41, 77–81, 111,
118, 192, 240
British Empire and, 84–85
law of gravity of, 5, 37, 145–46,
150–51, *152,* 245, 246

Nine Chains to the Moon, 8, 76, 79–80, 114, 255
Nonsimultaneousness, 105
Norquist, (president of grain bin company), 235–36
Novae, 238, 245
"Novent," definition of, 237n
Nuclear structural systems, *171*
Nucleus
  closest packing around, *164–66*
  coherence of, 238
  omnicongruence of, 244–45, *245*
Numbers
  prime
    "bad luck," 69
    eternal disquietude imposed by, 86
    for polyvertexia, *132*
    in synergetics' hierarchy, *128–29*
  "sixtiness" concept of, 92
  whole
    calculating only with, 68–70, 224
    nature works only with, 68
    in organic chemistry, 220

Octahedron, 41, 45, 50, 138. *See also* Coupler
  angular topology of, *198*
  as conservation and annihilation model, 247
  constellation of, 153–54, *153, 154*
  in filling allspace, 53
  loss of concept of, 97
  new name for, *132,* 148
  spherical, 190, 191, 235
  strength of, 222, *223*
  in synergetics' hierarchy, *128–29*
  in synergetic mathematics, 224
  tensegrity, *179*
  tetrahedral mensuration applied to, *199*
Ohm's law, 40, 235
One-dimensionality, 140, 169
Operational procedures, 30–31
Orbital rotation, 18, 208, *208–9. See also* Planets—orbits of
Otherness, 70, 124, 135
Oxygen, identification of, 35

Parameters, 127
Particle discontinuity vs. wave continuity, 194–95, 235

Patent claims, 76–77
Peace, artifact revolution's contribution to, 3
Peashooters, 20–22, *21*
Persia, 94
Philolaus, 100
Phoenicians, 94
Photons, 105, 106, 233
Photosynthesis, 241
  as syntropic, 73
Physics
  awkward cubic standard of, 52–53, 60, 129
  definition of, 131
  divorced from geometry, 46, 139–40
  particle, 63
  solid-state, 66
Pi, 70, 169, 192–93, *196*
Planck's Constant, 169, 228, 231
Planets
  human observation of, 35–36
  orbits of, 4–5, 20, 36, 82, 100, *208–9*
  Universe's creation of, 245
Plastic products, 114
Plato, 100, 130, 174, 220, 221
Pluto (planet), 82
Point. *See also* Vertexes (corners)
  cannot be passed through, by more than one line, 10–13, 119, *120,* 233
  not more than one event can occupy, 232
  synergetic connecting of point to, *242*
Polaris (North Star), 38, 105
Polarity, as additive twoness, 136–37, *137*
Polyhedra. *See also specific types of polyhedra*
  areal and volumetric domains of, 59–60n
  Euler's formula for, 58n, 122
  made of cheese, 228, *229*
  naming of, 130
  stones (rocks) as, 122, *123,* 138
  tetrahedral mensuration applied to, *199*
Polyvertexia
  definition of, 131
  high-frequency, 230, 233
  new identification of, *132*
  thinking only in terms of, 163

Population, Malthusian view of, 107–9, 111–12, 263
Power structure, 2, 260
  author's strategy toward, 7, 8
  catering to want and suffering by, 248–49
  democracy bought out by, 262
  divide-to-conquer strategy of, 250–51
  education influenced by, 37, 103, 249
  error-making forbidden by, 261–62
  traditional, soon to be obsolescent, 259, 264
Precession, 18–19, 24–27, 35, 208, 209–13, 240
  in social sphere, 27
  tetrahedral, of closest-packed spheres, *214, 215*
Presize, 50, 51
Pretime, 50, 51
Priestley, Joseph, 34
Primitive, use of word, 221
Principles
  discovered solely by human minds, 237
  as weightless, 218
Prism, *48, 120, 168*
  tetrahedral mensuration applied to, *199*
Propeller blades, 10–12, 65–66
Protoplasm, 39
Pythagoras, 100

Quadrangular accounting, 226, *227*
Quanta Modules
  A, 49, *54,* 55–56, *172,* 231, 240, 241, 244
    in coupler, 61–62
    definition of, 55*n*
    number of, 55
    time and, 57
  B, *54,* 55–56, 231, 241, 244
    in coupler, 61–62
    number of, 55
  C, D, E, F, 241
  predications of, 72–73
Quantum mechanics, 18–19, 24, 251–52
  multiplication by division in, 51–52, 52*n,* 69, 238
Quarks (mites), 58, 61–62, 232
Quasicube, 156–57, *156, 157*
Quasisphere, 196

Race schizophrenia, 81, *82*
Radiation
  all energies transformed into, 241
  coexistence of gravity and, 39, 51, 56–57, 89–90, *90, 200*
  as disintegration, 56
  same velocity of any type of, 105
Radio, 255–56
Rainbow, colors of, 12
Randomness, seeming. *See* Interrelationships
Reality
  invisible technological, 109–11
  minimum demonstrable, 239
  as spiro-orbital, *208–9*
Recycling of metals, 115–16
Reflection, *120*
Refraction, *120*
Regular, meaning of, in geometry, 169
Relative frequency, 13
Relative size advantage, 13–14, 238–40
Religions, 2, 248. *See also* Christianity; God; Roman Catholicism
Resistance, 235
Resonance, 86
Rocks. *See* Stones
Roebling, W. A., 110
Roemer, Olaus, 80, 84, 118
Roman Catholicism
  Earth as center of Universe in, 90, 100–101, 103
  emperor-popes of, 85, 101, 103
Roman Empire, 95, 96, 107
Roman numerals, 101–2
Roosevelt, Franklin D., 79

Scheherazade Number, 68–70
Science
  definition of, 131
  should be based on scientists' experimental evidence, 259
Scissors, *45, 46,* 47
Sensorial explanations, 19, 24–27
Seven Mysteries of Life, The (Murchie), 13, 238, 241
Shipbuilding
  ancient, 94–95
  of clipper ships, 13–14, 239
Shunting, 17–18, 67–68
Similitude, 13–14, 17, 39, 64, 238–40
Six-dimensionality, 144, *145*
Six-vertexion. *See* Octahedron
Sleep, 57

Smalley, Richard, 67*n*
Snyder-Fuller interattraction law, *152,*
   163
Solenoids, 235
Solidity, mistaken notion of, 12–13,
   65–66, 104, 130–31, 159, 174,
   178–80, 187, *188*
Sound, speed of, 256
Southeast Asia, ancient, 94
Space stations, 184
Space travel, 74
Specialization
   educational system's cultivation of,
     249
   self-perpetuating disease of, 251
   synergy as antithetical to, 32
Sperry, Elmer, 35
Sperry Gyroscope Company, 19
Spheres
   closest-packed, 58, *58, 142,* 159–63,
     164–67, 170, *170,* 230–32
     tetrahedral precession of, *214,*
       215
   contain most volume with least sur-
     face, 239
   definition of, 187–89, 192, 196,
     217
   energetic-effectiveness of increase in
     size of, 239
   Greek, 187, 213, 217
   impossibility of "cubing," 157–59,
     159
   new names for, *132*
   rhombic dodecahedra as domains of,
     59
   in synergetics' hierarchy, *128–29*
   tensegrity. *See* Tensegrity spheres
   vertexial, trivalent bonding of, *168*
Spherical excess, 190
Spherics, 160–63, *162*
Square
   in conventional geometry, 224–25
   Euclidean definition of, 99
Squaring, "triangling" preferred to,
   225–26, *227*
Standard of living. *See* Living standard
Stars
   as atomic energy plants, 260
   cosmic clouds and, 245–47
   interrelationships of, *132*
   as predominantly entropic systems,
     73
Sticks, falling, *201*

Stone masonry, tension and compres-
   sion in, 87, 109
Stones (rocks), as polyhedra, 122, 138,
   144
Stone-skipping, 121
Strassmann, Fritz, 79
Structure, 41–51. *See also* System
   definition of, 43
   three primitive systems of, 241–42
Struts
   cube stabilized by, 222, *223*
   tetrahedron masts as, *188,* 244
   triangle as basis of 50, 225–26
Subatomic particles, in synergetics,
   61–64
Sun, 261
   orbits of planets around, 4–5, 20, 36,
     82, 100, *208–9*
Supply-side economists, 29
Suspension bridges, *158*
Synergetics, 18–19, 31, 46*n,* 47, 66
   axioms lacking in, 119
   constants of hierarchy in, *128–29,*
     240–41
   conversion of conventional geometry
     to, 223
   Euclidean geometry and, 98
   multiplication by division in, 52*n,* 69
   particle physics and, 63–64
   simple models and whole numbers
     used by, 63
   tetrahedron's edge as unity in, 60
   unique advantages in using, 66–68,
     118, 220, 233, 245–47, 264
   Universe when divided by, 72
Synergy, definition of, 32
Syntropic, definition of, 9*n*
Syntropy, 73, 245
   definition of, 51
   quanta modules and, 56–57
System. *See also* Structure
   definition of, 124, 131, *134*
   degrees of freedom in, 131–34, 182–
     84, *183*
   micro-, 135, *136*
   mini-, 135, *136*
   minimum, 41, 45, 47, 50, 126, *126,*
     131, *134,* 135–36, 138, 169, 191,
     201, 228
     defined by four vectors of re-
       straint, *202*
   no compressional component of,
     touching another, 242

System (*cont.*)
 no parts independent of, 140–41
 realization of, 191
 sum of angles around, 191–93
Systemic thinking, 72, 91–92, 124–30, *126*

Tangential avoidance, *120*
Technology
 invisible revolution in, 109–11
 living standard advanced by, 262
 Malthusianism and, 108, 111–12
Tennis, 20
Tensegrity, 173–82, 232, 242
 adopted by dictionaries, 174*n*
 basic structures of, *179*
 definition of, 42*n*, 196
 in every geometrical structure, 187
Tensegrity masts, 184–87, *185, 186, 188*, 244
Tensegrity spheres, 87–90, *175, 176*
 balloon as example of, 174–82, *181*
 basic principles of, 180–81, *181*
 single and double bonding in, *176*
Tensional integrity. *See* Tensegrity
Tension and compression, 87–88, 109, 234
Tetrahedron (tetravertexion; four-vertexion), 41, *42, 43*, 45–47, *48*
 angular topology of, *198*
 conceptual defining of Universe by, 71–72
 as conservation and annihilation model, 244
 constant properties of, 229, 241, *242*
 constellations of, 148–59, *149, 152–55*
 coordination system based on, 52–55, *128–29*
 division of, 169, *172*
 four-dimensionality of, 230
 four spheres locking as, *162*
 historical loss of concept of, 97, 220
 inside-outing of, *43, 49, 49*, 57, 205
 made of cheese, 228, *229*
 as minimum structural system of Universe, 41, 45, 47, 50, 126, *126*, 131, *134*, 135–36, 138, 169, 191, *201, 216*, 228
 new name for, *132*
 in nuclear structural systems, *171*
 six great-circle-spun subdivisions of, *54*, 55

spherical, 190
stone transforms to, 122, *123*
sum of angles around vertexes of, 191
symmetry of, 169, 229
in synergetic mathematics, 60, 224, 225
tensegrity, *179*, 184–87, *186*, 244
three ways of demonstrating, *142*
truncated, 141, *143, 171*
Tetrakytis, 230
Tetrascroll (book), 27
Tetravertexion. *See* Tetrahedron
Thought systems, geometry of, 124, 130, 138
Three-dimensionality, 140, 169
Three Mile Island incident, 257
Tides, 37
Time, 50, 57
 Newton's assumption about, 77, 79–80
 relativity of, 77
Titanium, 174
Topology, 39, 118, 122, 131, 141–44, 217–18
 angular, *198*
 definition of, 141*n*
Torque, 18
 minimum of twelve spokes to oppose, *204*
Transistors, 66
Triacontahedron
 rhombic, new name for, *132*
 tetrahedral mensuration applied to, *199*
Triangle
 adjustable spherical, 189–91, *189*
 angular topology of, *198*
 as basis of structural systems in Universe, 50, 225–26
 constancy of, 229
 Euclidean definition of, 99
 Greek definition of, 193
 inside-outing of, *206*
 right, 149–50, *150*
 spherical, 193–94, *194, 195*
 stability of, 42–47, *44, 45, 48*, 241
 sum of angles of, 194, *194*
 three-way grid of great circles in, 236
Trigonometric functions, synergetic reworking of calculations of, 69
Trigonometry, 34, 84, 98
 spherical, 187–97, 200, 204, 238
Troy, 94–95

Tuning, definition of, 235
Two-dimensionality, 140, 143, 169
Twoness, 48–49, 57–58, 136–37, *137,
189*, 208, 217, 237

Unified field, 39, 42*n*, 51, 89–90, *90,*
174
Unity
as plural, 48–49, 57–58, 208
tetrahedron's vs. cube's edge as, 60
Universal joint, *204*
Universe
definition of, 64, 236
degrees of freedom in, 19, 182–84,
183, 199–205, *204, 242*
Einstein's concept of, distinguished
from Newton's, 38, 80–86, 106
Einstein's description of, 40, 80–86,
105–6
as eternal, 38, 50, 64
eternal regeneration of, 51, 57, 64,
82–83, 105, 241, 247, 259
as finite aggregate of finites, 105
generalized principles (laws) of, 3,
5–6, 10, 29–30, 33, 38–40, 64, 118
human function in, 2, 3–4, 6–7, 40,
116
inherent twoness of, 48–49, 57–58,
217, 237
learning must start with, 39
no beginning or ending to, 38, 40, 84
nothing is lost in, 106
no touching of two things in, 130,
145, 163, 174, 178–80
no *up* or *down* in, 82, 104
only three primitive structural systems in, 241–42
orderliness of, as cosmic religious
sense, 75–76, 259
as resonant, 86
Rome as center of, 100–101
scenario, 38, 40, 64, 72–73, 105–6
definition of, 236
shortest routes through, 235
six fundamental motions of, 199–205
spherical islands of, 88
systemic division of, 72, 91–92,
124–27, *126*
as tensegrity, 173
tension and compression in, 88,
195–96
tetrahedron as minimum structural
system of, 41, 45, 47, 50, 126,

126, 131, *134*, 135–36, 138, 169,
191, *201, 216*, 228
whole-number accounting in, 68–70
Up and *down*, 82, 104

van't Hoff, Jacobus H., 45, 220
Vector equilibrium (VE), 67, *132, 164,*
208–9, *210*, 240
angular topology of, *198*
constructed from foldable great circles, 234–35, *234*
"jitterbug" transformation of, 230,
231
in synergetic mathematics, 224
Vector matrix, isotropic, 53–54, 58,
60, 128, 169, *171*, 230–31
Vectors
of falling sticks, *201*
gravity and radiation, 89–90, *90*
push-pull, 161
of restraint, as defining minimum
system, *202*
use of term in synergetics, 45*n,*
142
Venice, 95
Vertexes (corners), 141, *142. See also*
Euler's formula; Flex-corners

War
elimination of, 3
huge expenditures on, 109, 257
technology of and growth of life-
supporting technology, 17–18, 28,
29, 113, 115
Wave continuity vs. particle discontinu-
ity, 194–95, 235
Weiss, Paul, 65–66
Wiener, Norbert, 74
Winds, blowing of, 104–5
Wire wheel, 88
Wisdom, 39
Wood, tension and compression in, 87–
88, 109
World War I, 110, 111
World War II, 110

X-ray diffraction gratings, 11–12
XYZ frame of reference, 63, 99, 140,
196, 197, 220, 228

Yale University, 65
Yin-yang, *138*
Young people, 256–58